Applied and Computational Statistics

Applied and Computational Statistics

Special Issue Editor
Sorana D. Bolboacă

MDPI • Basel • Beijing • Wuhan • Barcelona • Belgrade

Special Issue Editor
Sorana D. Bolboacă
Iuliu Hațieganu University of Medicine and Pharmacy
Romania

Editorial Office
MDPI
St. Alban-Anlage 66
4052 Basel, Switzerland

This is a reprint of articles from the Special Issue published online in the open access journal *Mathematics* (ISSN 2227-7390) from 2018 to 2019 (available at: https://www.mdpi.com/journal/mathematics/special_issues/applied_computational_statistics).

For citation purposes, cite each article independently as indicated on the article page online and as indicated below:

LastName, A.A.; LastName, B.B.; LastName, C.C. Article Title. *Journal Name* **Year**, *Article Number*, Page Range.

ISBN 978-3-03928-176-3 (Pbk)
ISBN 978-3-03928-177-0 (PDF)

© 2020 by the authors. Articles in this book are Open Access and distributed under the Creative Commons Attribution (CC BY) license, which allows users to download, copy and build upon published articles, as long as the author and publisher are properly credited, which ensures maximum dissemination and a wider impact of our publications.

The book as a whole is distributed by MDPI under the terms and conditions of the Creative Commons license CC BY-NC-ND.

Contents

About the Special Issue Editor vii

Preface to "Applied and Computational Statistics" ix

Ibrahim Elbatal, Farrukh Jamal, Christophe Chesneau, Mohammed Elgarhy and Sharifah Alrajhi
The Modified Beta Gompertz Distribution: Theory and Applications
Reprinted from: *Mathematics* **2019**, *7*, 3, doi:10.3390/math7010003 1

Lorentz Jäntschi and Sorana D. Bolboacă
Computation of Probability Associated with Anderson–Darling Statistic
Reprinted from: *Mathematics* **2018**, *6*, 88, doi:10.3390/math6060088 18

Miltiadis S. Chalikias
Optimal Repeated Measurements for Two Treatment Designs with Dependent Observations: The Case of Compound Symmetry
Reprinted from: *Mathematics* **2019**, *7*, 378, doi:10.3390/math7040378 35

Lili Tan, Yunzhan Gong and Yawen Wang
A Model for Predicting Statement Mutation Scores
Reprinted from: *Mathematics* **2019**, *7*, 778, doi:10.3390/math7090778 41

Dan-Marian Joița and Lorentz Jäntschi
Extending the Characteristic Polynomial for Characterization of C_{20} Fullerene Congeners
Reprinted from: *Mathematics* **2017**, *5*, 84, doi:10.3390/math5040084 80

About the Special Issue Editor

Sorana D. Bolboacă is a professor of medical informatics and biostatistics at the "Iuliu Hațieganu" University of Medicine and Pharmacy Cluj-Napoca, Romania. She earned her Ph.D. in Medicine (2006) from the Iuliu Hațieganu University of Medicine and Pharmacy (title of the thesis *Evidence-Based Medicine: Logistics and Implementation*) and a Ph.D. in Horticulture (2010) from the University of Agriculture Sciences and Veterinary Medicine Cluj-Napoca (title of the thesis *Statistical Models for Analysis of Genetic Variability*). Her research interests are multidisciplinary, and include applied and computational statistics, molecular modeling, genetic analysis, statistical modeling in medicine, integrated health informatics systems and the application of new technologies in medicine, medical diagnostics research, medical imaging analysis, assisted decision systems, research ethics, social media and health information, and evidence-based medicine. She is the author of more than 200 papers and 19 monographs in medicine, computational chemistry, computer science, mathematics, environmental sciences, biomedical engineering, nanoscience nanotechnology, and medical informatics.

Preface to "Applied and Computational Statistics"

The research on statistical populations, samples, or models have applications in all research areas and are conducted to gain knowledge for real-world problems. Increased calculation power opens the path to computational statistics, algorithm translation, and the implementation of statistical methods and computer simulations. These areas are developing rapidly, providing solutions to multidisciplinary, interdisciplinary, and transdisciplinary topics. An excellent theoretical statistics method is worthless in the absence of real-data applicability. Furthermore, any excellent theoretical statistics method will find its ending sooner or later without proper implementation.

Statistical methods find their application is understanding phenomenon from all fields, including medicine, biology, biochemistry, agriculture, horticulture, engineering, and more. The Special Issue of *Mathematics* entitled "Applied and Computational Statistics" provides new methods and their applicability to the prediction of the mutation score, repeated measurements bi-treatment cross-over design, modified beta Gompertz distribution, extending the characteristic polynomial, and computation of the probability associated with Anderson–Darling statistics. In addition to giving a detailed presentation of the implemented method that assures the reproducibility, the collection of articles also includes specific applicability examples.

Thanks to all contributors for their involvement in finding statistical solutions to real-life problems; keep staying on the path of knowledge gain. Dear reviewers, thank you very much for the time spent in reviewing the articles and for the constructive comments that have certainly contributed to the quality of the papers. I warmly invite readers to enjoy reading this collection of articles, and I hope that new ideas will come to life for the benefit of science by reading these manuscripts.

Sorana D. Bolboacă
Special Issue Editor

Article

The Modified Beta Gompertz Distribution: Theory and Applications

Ibrahim Elbatal [1], Farrukh Jamal [2], Christophe Chesneau [3,*], Mohammed Elgarhy [4] and Sharifah Alrajhi [5]

1. Institute of Statistical Studies and Research (ISSR), Department of Mathematical Statistics, Cairo University, Giza 12613, Egypt; i_elbatal@staff.cu.edu.eg
2. Department of Statistics, Govt. S.A Postgraduate College Dera Nawab Sahib, Bahawalpur, Punjab 63360, Pakistan; drfarrukh1982@gmail.com
3. Department of Mathematics, LMNO, University of Caen, 14032 Caen, France
4. Department of Statistics, University of Jeddah, Jeddah 21589, Saudi Arabia; m_elgarhy85@yahoo.com
5. Department of Statistics, Faculty of Sciences, King Abdulaziz University, Jeddah 21589, Saudi Arabia; saalrajhi@kau.edu.sa
* Correspondence: christophe.chesneau@unicaen.fr; Tel.: +33-02-3156-7424

Received: 7 November 2018; Accepted: 17 December 2018; Published: 20 December 2018

Abstract: In this paper, we introduce a new continuous probability distribution with five parameters called the modified beta Gompertz distribution. It is derived from the modified beta generator proposed by Nadarajah, Teimouri and Shih (2014) and the Gompertz distribution. By investigating its mathematical and practical aspects, we prove that it is quite flexible and can be used effectively in modeling a wide variety of real phenomena. Among others, we provide useful expansions of crucial functions, quantile function, moments, incomplete moments, moment generating function, entropies and order statistics. We explore the estimation of the model parameters by the obtained maximum likelihood method. We also present a simulation study testing the validity of maximum likelihood estimators. Finally, we illustrate the flexibility of the distribution by the consideration of two real datasets.

Keywords: modified beta generator; gompertz distribution; maximum likelihood estimation

MSC: 60E05; 62E15; 62F10

---

1. Introduction

The Gompertz distribution is a continuous probability distribution introduced by Gompertz [1]. The literature about the use of the Gompertz distribution in applied areas is enormous. A nice review can be found in [2], and the references therein. From a mathematical point of view, the cumulative probability density function (cdf) of the Gompertz distribution with parameters $\lambda > 0$ and $\alpha > 0$ is given by

$$G(x) = 1 - e^{-\frac{\lambda}{\alpha}(e^{\alpha x}-1)}, \quad x > 0.$$

The related probability density function (pdf) is given by

$$g(x) = \lambda e^{\alpha x} e^{-\frac{\lambda}{\alpha}(e^{\alpha x}-1)}, \quad x > 0.$$

It can be viewed as a generalization of the exponential distribution (obtained with $\alpha \to 0$) and thus an alternative to the gamma or Weibull distribution. A feature of the Gompertz distribution is that $g(x)$ is unimodal and has positive skewness, whereas the related hazard rate function (hrf) given by $h(x) = g(x)/(1-G(x))$ is increasing. To increase the flexibility of the Gompertz distribution, further extensions have been proposed. A natural one is the generalized Gompertz distribution introduced by El-Gohary et al. [3]. By introducing an exponent parameter $a > 0$, the related cdf is given by

$$F(x) = \left(1 - e^{-\frac{\lambda}{\alpha}(e^{\alpha x}-1)}\right)^a, \quad x > 0.$$

The related applications show that a plays an important role in term of model flexibility. This idea was then extended by Jafari et al. [4] who used the so-called beta generator introduced by Eugene et al. [5]. The related cdf is given by

$$\begin{aligned} F(x) &= \frac{1}{B(a,b)} \int_0^{1-e^{-\frac{\lambda}{\alpha}(e^{\alpha x}-1)}} t^{a-1}(1-t)^{b-1} dt \\ &= I_{1-e^{-\frac{\lambda}{\alpha}(e^{\alpha x}-1)}}(a,b), \quad x > 0. \end{aligned} \quad (1)$$

where $a, b > 0$, $B(a,b) = \int_0^1 t^{a-1}(1-t)^{b-1} dt$ and $I_x(a,b) = (1/B(a,b)) \int_0^x t^{a-1}(1-t)^{b-1} dt$, $x \in [0,1]$. This distribution has been recently extended by Benkhelifa [6] with a five-parameter distribution. It is based on the beta generator and the generalized Gompertz distribution.

Motivated by the emergence of complex data from many applied areas, other extended Gompertz distributions have been proposed in the literature. See for instance, El-Damcese et al. [7] who considered the Odd Generalized Exponential generator introduced by Tahir et al. [8]; Roozegar et al. [9] who used the McDonald generator introduced by Alexander et al. [10]; Refs. [11,12] who applied the transmuted generator introduced by Shaw and Buckley [13]; Chukwu and Ogunde [14] and Lima et al. [15] who used the Kumaraswamy generator; and Benkhelifa [16] and Yaghoobzadeh [17] who considered the Marshall–Olkin generator introduced by Marshall and Olkin [18]; and Shadrokh and Yaghoobzadeh [19] who considered the Beta-G and Geometric generators.

In this paper, we present and study a distribution with five parameters extending the Gompertz distribution. It is based on the modified beta generator developed by Nadarajah et al. [20] (which can also be viewed as a modification of the beta Marshall–Olkin generator developed by Alizadeh et al. [21]). The advantage of this generator is to nicely combine the advantages of the beta generator of Eugene et al. [5] and the Marshall–Olkin generator of Marshall and Olkin [18]. To the best of our knowledge, its application to the Gompertz distribution has never been considered before. We provide a comprehensive description of its general mathematical properties (expansions of the cdf and pdf, quantile function, various kinds of moments, moment generating function, entropies and order statistics). The estimation of the model parameters by maximum likelihood is then discussed. Finally, we explore applications to real datasets that illustrate the usefulness of the proposed model.

The structure of the paper is as follows. Section 2 describes the considered modified beta Gompertz distribution. Some mathematical properties are investigated in Section 3. Section 4 provides the necessary to the estimation of the unknown parameters with the maximum likelihood method. A simulation study is presented, which tests validity of the obtained maximum likelihood estimators. Applications to two real datasets are also given.

2. The Modified Beta Gompertz Distribution

Let $c > 0$, $G(x)$ be a cdf and $g(x)$ be a related pdf. The modified beta generator introduced by Nadarajah et al. [20] is characterized by the cdf given by

$$F(x) = I_{\frac{cG(x)}{1-(1-c)G(x)}}(a,b), \quad (2)$$

By differentiation of $F(x)$, a pdf is given by

$$f(x) = \frac{c^a g(x) [G(x)]^{a-1} [1-G(x)]^{b-1}}{B(a,b)[1-(1-c)G(x)]^{a+b}}, \quad x \in \mathbb{R}. \tag{3}$$

The hrf is given by

$$h(x) = \frac{c^a g(x) [G(x)]^{a-1} [1-G(x)]^{b-1}}{B(a,b)[1-(1-c)G(x)]^{a+b}\left(1 - I_{\frac{cG(x)}{1-(1-c)G(x)}}(a,b)\right)}, \quad x \in \mathbb{R}.$$

Let us now present our main distribution of interest. Using the cdf $G(x)$ of the Gompertz distribution with parameters $\lambda > 0$ and $\alpha > 0$ as baseline, the cdf given by Equation (2) becomes

$$F(x) = I_{\frac{c\left(1-e^{-\frac{\lambda}{\alpha}(e^{\alpha x}-1)}\right)}{1-(1-c)\left(1-e^{-\frac{\lambda}{\alpha}(e^{\alpha x}-1)}\right)}}(a,b), \quad x > 0. \tag{4}$$

The related distribution is called the modified beta Gompertz distribution (MBGz distribution), also denoted by MBGz$(\lambda, \alpha, a, b, c)$. The related pdf in Equation (3) is given by

$$f(x) = \frac{c^a \lambda e^{\alpha x} e^{-\frac{\lambda b}{\alpha}(e^{\alpha x}-1)} \left(1 - e^{-\frac{\lambda}{\alpha}(e^{\alpha x}-1)}\right)^{a-1}}{B(a,b)\left[1-(1-c)\left(1-e^{-\frac{\lambda}{\alpha}(e^{\alpha x}-1)}\right)\right]^{a+b}}, \quad x > 0. \tag{5}$$

The hrf is given by

$$h(x) = \frac{c^a \lambda e^{\alpha x} e^{-\frac{\lambda b}{\alpha}(e^{\alpha x}-1)} \left(1 - e^{-\frac{\lambda}{\alpha}(e^{\alpha x}-1)}\right)^{a-1}}{B(a,b)\left[1-(1-c)\left(1-e^{-\frac{\lambda}{\alpha}(e^{\alpha x}-1)}\right)\right]^{a+b}\left[1 - I_{\frac{c\left(1-e^{-\frac{\lambda}{\alpha}(e^{\alpha x}-1)}\right)}{1-(1-c)\left(1-e^{-\frac{\lambda}{\alpha}(e^{\alpha x}-1)}\right)}}(a,b)\right]},$$

$x > 0$. $\tag{6}$

Figure 1 shows the plots for $f(x)$ and $h(x)$ for selected parameter values λ, α, a, b, c. We observe that these functions can take various curvature forms depending on the parameter values, showing the increasing of the flexibility of the former Gompertz distribution.

A strong point of the MBGz distribution is to contain different useful distributions in the literature. The most popular of them are listed below.

- When $c = 1/(1-\theta)$ with $\theta \in (0,1)$ (θ is a proportion parameter), we obtain the beta Gompertz geometric distribution introduced by Shadrokh and Yaghoobzadeh [19], i.e., with cdf

$$F(x) = I_{\frac{1-e^{-\frac{\lambda}{\alpha}(e^{\alpha x}-1)}}{1-\theta e^{-\frac{\lambda}{\alpha}(e^{\alpha x}-1)}}}(a,b), \quad x > 0.$$

However, this distribution excludes the case $c \in (0,1)$, which is of importance since it contains well-known flexible distributions, as developed below. Moreover, the importance of small values for c can also be determinant in the applications (see Section 4).

- When $c = 1$, we get the beta Gompertz distribution with four parameters introduced by Jafari et al. [4], i.e., with cdf

$$F(x) = I_{1-e^{-\frac{\lambda}{\alpha}(e^{\alpha x}-1)}}(a,b), \quad x > 0.$$

- When $c = b = 1$, we get the generalized Gompertz distribution studied by El-Gohary et al. [3], i.e., with cdf

$$F(x) = \left(1 - e^{-\frac{\lambda}{\alpha}(e^{\alpha x}-1)}\right)^a, \quad x > 0.$$

- When $a = b = 1$ and $c = \frac{1}{\theta}$ with $\theta > 1$, we get the a particular case of the Marshall–Olkin extended generalized Gompertz distribution introduced by Benkhelifa [16], i.e., with cdf

$$F(x) = \frac{1 - e^{-\frac{\lambda}{\alpha}(e^{\alpha x}-1)}}{\theta + (1-\theta)\left(1 - e^{-\frac{\lambda}{\alpha}(e^{\alpha x}-1)}\right)}, \quad x > 0.$$

- When $a = b = c = 1$, we get the Gompertz distribution introduced by Gompertz [1], i.e., with cdf

$$F(x) = 1 - e^{-\frac{\lambda}{\alpha}(e^{\alpha x}-1)}, \quad x > 0.$$

- When $c = 1$ and $\alpha \to 0$, we get beta exponential distribution studied by Nadarajah and Kotz [22], i.e., with cdf

$$F(x) = I_{1-e^{-\lambda x}}(a,b), \quad x > 0.$$

- When $b = c = 1$ and $\alpha \to 0$, we get the generalized exponential distribution studied by Gupta and Kundu [23], i.e., with cdf

$$F(x) = \left(1 - e^{-\lambda x}\right)^a, \quad x > 0.$$

- When $a = b = c = 1$ and $\alpha \to 0$ we get the exponential distribution, i.e., with cdf

$$F(x) = 1 - e^{-\lambda x}, \quad x > 0.$$

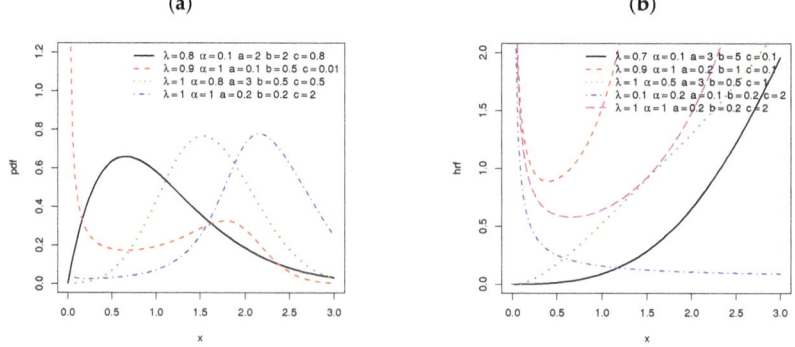

Figure 1. Some plots of the pdf $f(x)$ (**a**); and some plots for the hrf $h(x)$ (**b**).

3. Some Mathematical Properties

3.1. On the Shapes of the pdf

The shapes of $f(x)$ given by Equation (5) can be described analytically. As usual, the critical points x_* of the pdf $f(x)$ satisfies $\frac{\partial}{\partial x} \ln(f(x_*)) = 0$, with

$$\frac{\partial}{\partial x} \ln(f(x)) = \alpha - b\lambda e^{\alpha x} + (a-1)\frac{\lambda e^{\alpha x} e^{-\frac{\lambda}{\alpha}(e^{\alpha x}-1)}}{1 - e^{-\frac{\lambda}{\alpha}(e^{\alpha x}-1)}} + (a+b)(1-c)\frac{\lambda e^{\alpha x} e^{-\frac{\lambda}{\alpha}(e^{\alpha x}-1)}}{c + (1-c)e^{-\frac{\lambda}{\alpha}(e^{\alpha x}-1)}}.$$

The point x_* is a local maximum if $\frac{\partial^2}{\partial x^2} \ln(f(x_*)) < 0$, a local minimum if $\frac{\partial^2}{\partial x^2} \ln(f(x_*)) > 0$ and a point of inflection if $\frac{\partial^2}{\partial x^2} \ln(f(x_*)) = 0$.

Let us now study the asymptotic properties of $f(x)$. We have

$$f(x) \sim \frac{c^a}{B(a,b)} \lambda^a x^{a-1}, \quad x \to 0.$$

Thus, for $a \in (0,1)$, we have $\lim_{x \to 0} f(x) = +\infty$; for $a = 1$, we have $\lim_{x \to 0} f(x) = bc\lambda$; and, for $a > 1$, we have $\lim_{x \to 0} f(x) = 0$. We have

$$f(x) \sim \frac{1}{c^b B(a,b)} \lambda e^{\alpha x} e^{\frac{\lambda b}{\alpha}} e^{-\frac{\lambda b}{\alpha} e^{\alpha x}}, \quad x \to +\infty.$$

Thus, $\lim_{x \to +\infty} f(x) = 0$ in all cases. Figure 1a illustrates these points for selected parameters.

3.2. On the Shapes of the hrf

Similar to the pdf, the critical points x_* of the hrf $h(x)$ given by Equation (6) satisfy $\frac{\partial}{\partial x} \ln(h(x_*)) = 0$, with

$$\frac{\partial}{\partial x} \ln(h(x)) = \alpha - b\lambda e^{\alpha x} + (a-1)\frac{\lambda e^{\alpha x} e^{-\frac{\lambda}{\alpha}(e^{\alpha x}-1)}}{1 - e^{-\frac{\lambda}{\alpha}(e^{\alpha x}-1)}} + (a+b)(1-c)\frac{\lambda e^{\alpha x} e^{-\frac{\lambda}{\alpha}(e^{\alpha x}-1)}}{c + (1-c)e^{-\frac{\lambda}{\alpha}(e^{\alpha x}-1)}}$$

$$+ \frac{c^a \lambda e^{\alpha x} e^{-\frac{\lambda b}{\alpha}(e^{\alpha x}-1)} \left(1 - e^{-\frac{\lambda}{\alpha}(e^{\alpha x}-1)}\right)^{a-1}}{B(a,b)\left[1 - (1-c)\left(1 - e^{-\frac{\lambda}{\alpha}(e^{\alpha x}-1)}\right)\right]^{a+b}\left[1 - I_{\frac{c\left(1-e^{-\frac{\lambda}{\alpha}(e^{\alpha x}-1)}\right)}{1-(1-c)\left(1-e^{-\frac{\lambda}{\alpha}(e^{\alpha x}-1)}\right)}}(a,b)\right]}.$$

Again, the point x_* is a local maximum if $\frac{\partial^2}{\partial x^2} \ln(h(x_*)) < 0$, a local minimum if $\frac{\partial^2}{\partial x^2} \ln(h(x_*)) > 0$ and a point of inflection if $\frac{\partial^2}{\partial x^2} \ln(h(x_*)) = 0$.

We also have

$$h(x) \sim \frac{c^a}{B(a,b)} \lambda^a x^{a-1}, \quad x \to 0.$$

Thus, for $a \in (0,1)$, we have $\lim_{x \to 0} h(x) = +\infty$; for $a = 1$, we have $\lim_{x \to 0} h(x) = bc\lambda$; and, for $a > 1$, we have $\lim_{x \to 0} h(x) = 0$. We have

$$h(x) \sim b\lambda e^{\alpha x}, \quad x \to +\infty.$$

Thus, $\lim_{x \to +\infty} h(x) = +\infty$ in all cases. Figure 1b illustrates these points for selected parameters.

3.3. Linear Representation

Let us determine useful linear representations for $F(x)$ given by Equation (4) and $f(x)$ given by Equation (5). First, let us suppose that $c \in (0,1)$. It follows from the generalized binomial formula, i.e., $(1+z)^\gamma = \sum_{k=0}^{+\infty} \binom{\gamma}{k} z^k$ for $|z| < 1$ and $\gamma \in \mathbb{R}$, with $\binom{\gamma}{k} = \frac{\gamma(\gamma-1)\ldots(\gamma-k+1)}{k!}$, that

$$F(x) = \frac{1}{B(a,b)} \int_0^{\frac{cG(x)}{1-(1-c)G(x)}} t^{a-1}(1-t)^{b-1} dt$$

$$= \frac{1}{B(a,b)} \sum_{k=0}^{+\infty} \binom{b-1}{k}(-1)^k \int_0^{\frac{cG(x)}{1-(1-c)G(x)}} t^{a+k-1} dt$$

$$= \frac{1}{B(a,b)} \sum_{k=0}^{+\infty} \binom{b-1}{k} \frac{(-1)^k}{a+k} \left[\frac{cG(x)}{1-(1-c)G(x)}\right]^{a+k}.$$

On the other hand, using again the generalized binomial formula, we obtain

$$\left[\frac{cG(x)}{1-(1-c)G(x)}\right]^{a+k} = c^{a+k} \sum_{\ell=0}^{+\infty} \binom{-(a+k)}{\ell}(-1)^\ell (1-c)^\ell [G(x)]^{\ell+a+k}. \tag{7}$$

In a similar manner, we have

$$[G(x)]^{\ell+a+k} = \left[1 - e^{-\frac{\lambda}{\alpha}(e^{\alpha x}-1)}\right]^{\ell+a+k} = \sum_{m=0}^{+\infty} \binom{\ell+a+k}{m}(-1)^m (1-H_m(x)),$$

where $H_m(x) = 1 - e^{-\frac{m\lambda}{\alpha}(e^{\alpha x}-1)}$ is the cdf of a Gompertz distribution with parameters $m\lambda$ and α. Combining these equalities, we obtain the following series expansion:

$$F(x) = \sum_{m=0}^{+\infty} v_m (1 - H_m(x)), \tag{8}$$

where

$$v_m = \frac{(-1)^m}{B(a,b)} \sum_{k=0}^{+\infty} \sum_{\ell=0}^{+\infty} \binom{\ell+a+k}{m} \binom{-(a+k)}{\ell} \binom{b-1}{k} c^{a+k}(-1)^{k+\ell}(1-c)^\ell \frac{1}{a+k}.$$

By derivation of $F(x)$, $f(x)$ can be expressed as

$$f(x) = \sum_{m=0}^{+\infty} w_m h_m(x), \tag{9}$$

where $w_m = -v_m$ and $h_m(x)$ is the pdf of a Gompertz distribution with parameters $m\lambda$ and α.

For the case $c > 1$, we must do some transformation for Equation (7) to apply the generalized binomial formula. We can write

$$\left[\frac{cG(x)}{1-(1-c)G(x)}\right]^{a+k} = \left[\frac{G(x)}{1-(1-\frac{1}{c})(1-G(x))}\right]^{a+k}$$

$$= [G(x)]^{a+k} \sum_{\ell=0}^{+\infty} \binom{-(a+k)}{\ell}(-1)^{\ell}(c-1)^{\ell}c^{-\ell}[1-G(x)]^{\ell}$$

$$= \sum_{\ell=0}^{+\infty}\sum_{q=0}^{\ell} \binom{-(a+k)}{\ell}\binom{\ell}{q}(-1)^{\ell+q}(c-1)^{\ell}c^{-\ell}[G(x)]^{q+a+k}.$$

On the other hand, we have

$$[G(x)]^{q+a+k} = \sum_{m=0}^{+\infty}\binom{q+a+k}{m}(-1)^{m}(1-H_m(x)).$$

Therefore, we can write $F(x)$ as Equation (8) with

$$v_m^* =$$

$$\frac{(-1)^m}{B(a,b)}\sum_{k=0}^{+\infty}\sum_{\ell=0}^{+\infty}\sum_{q=0}^{\ell}\binom{-(a+k)}{\ell}\binom{\ell}{q}\binom{q+a+k}{m}\binom{b-1}{k}(-1)^{\ell+q+k}(c-1)^{\ell}c^{-\ell}\frac{1}{a+k},$$

and $f(x)$ as Equation (9) with $w_m = -v_m^*$ (and still $h_m(x)$ is the pdf of a Gompertz distribution with parameters $m\lambda$ and α). For the sake of simplicity, we refer to the form in Equation (9) far all series representation of $f(x)$, whether $c \in (0,1)$ or $c > 1$.

Hereafter, we denote by X a random variable having the cdf $F(x)$ given by Equation (4) (and the pdf $f(x)$ given by Equation (5)) and by Y_m a random variable following the Gompertz distribution with parameters $m\lambda$ and α, i.e., having the cdf $H_m(x)$ (and the pdf $h_m(x)$).

3.4. Quantile Function

The quantile function of X is given by

$$Q(u) = \frac{1}{\alpha}\ln\left(1 - \frac{\alpha}{\lambda}\ln\left(1 - \frac{I_u^{-1}(a,b)}{c+(1-c)I_u^{-1}(a,b)}\right)\right), \quad u \in (0,1),$$

where $I_u^{-1}(a,b)$ denotes the inverse of $I_u(a,b)$. It satisfies $F(Q(u)) = Q(F(u)) = u$. Using [20], one can show that

$$Q(u) \sim \frac{1}{\lambda c}a^{\frac{1}{a}}B(a,b)^{\frac{1}{a}}u^{\frac{1}{a}}, \quad u \to 0.$$

From $Q(u)$, we can simulate the MBGz distribution. Indeed, let U be a random variable following the uniform distribution over $(0,1)$. Then, the random variable $X = Q(U)$ follows the MBGz distribution.

The median of X is given by $M = Q(1/2)$. We can also use $Q(u)$ to define skewness measures. Let us just introduce the Bowley skewness based on quartiles and the Moors kurtosis respectively defined by

$$B = \frac{Q(3/4)+Q(1/4)-2Q(1/2)}{Q(3/4)-Q(1/4)}, \quad M_o = \frac{Q(7/8)-Q(5/8)+Q(3/8)-Q(1/8)}{Q(6/8)-Q(2/8)}.$$

Contrary to γ_1 and γ_2, these quantities have the advantage to be always defined. We refer to [24,25].

3.5. Moments

Let r be a positive integer. The rth ordinary moment of X is defined by $\mu'_r = \mathbb{E}(X^r) = \int_{-\infty}^{+\infty} x^r f(x) dx$. Using the linear representation given by Equation (9), we can express μ'_r as

$$\mu'_r = \sum_{m=0}^{+\infty} w_m \int_{-\infty}^{+\infty} x^r h_m(x) dx = \sum_{m=0}^{+\infty} w_m \mathbb{E}(Y_m^r).$$

By doing the change of variables $u = e^{\alpha x}$, we obtain

$$\mathbb{E}(Y_m^r) = \frac{m\lambda}{\alpha^{r+1}} e^{\frac{m\lambda}{\alpha}} \int_1^{+\infty} (\ln u)^r e^{-\frac{m\lambda}{\alpha} u} du.$$

This integral has connections with the so-called generalized integro-exponential function. Further developments can be found in [26,27]. Therefore, we have

$$\mu'_r = \sum_{m=0}^{+\infty} w_m \frac{m\lambda}{\alpha^{r+1}} e^{\frac{m\lambda}{\alpha}} \int_1^{+\infty} (\ln u)^r e^{-\frac{m\lambda}{\alpha} u} du.$$

Obviously, the mean of X is given by $\mathbb{E}(X) = \mu'_1$ and the variance of X is given by $\mathbb{V}(X) = \mu'_2 - (\mu'_1)^2$.

3.6. Skewness

The rth central moment of X is given by $\mu_r = \mathbb{E}\left[(X - \mu'_1)^r\right]$. It follows from the binomial formula that

$$\mu_r = \sum_{k=0}^r \binom{r}{k} (-1)^k (\mu'_1)^k \mu'_{r-k}.$$

On the other side, the rth cumulants of X can be obtained via the equation:

$$\kappa_r = \mu'_r - \sum_{k=1}^{r-1} \binom{r-1}{k-1} \kappa_k \mu'_{r-k},$$

with $\kappa_1 = \mu'_1$. The skewness of X is given by $\gamma_1 = \kappa_3 / \kappa_2^{3/2}$ and the kurtosis of X is given by $\gamma_2 = \kappa_4 / \kappa_2^2$. One can also introduce the MacGillivray skewness given by

$$\rho(u) = \frac{Q(1-u) + Q(u) - 2Q(1/2)}{Q(1-u) - Q(u)}, \quad u \in (0,1).$$

It illustrates the effects of the parameters a, b, α and λ on the skewness. Further details can be found in [28].

3.7. Moment Generating Function

The moment generating function of X is given by $M_X(t) = \mathbb{E}\left(e^{tX}\right) = \int_{-\infty}^{+\infty} e^{tx} f(x) dx$. Using Equation (9), we have

$$M_X(t) = \sum_{m=0}^{+\infty} w_m \int_{-\infty}^{+\infty} e^{tx} h_m(x) dx = \sum_{m=0}^{+\infty} w_m M_{Y_m}(t),$$

where $M_{Y_m}(t) = \mathbb{E}(e^{tY_m})$, the moment generating function of Y_m. Doing successively the change of variables $u = e^{\alpha x}$ and the change of variable $v = \frac{m\lambda}{\alpha} u$, we obtain

$$M_{Y_m}(t) = \frac{m\lambda}{\alpha} e^{\frac{m\lambda}{\alpha}} \int_1^{+\infty} u^{\frac{t}{\alpha}} e^{-\frac{m\lambda}{\alpha} u} du = e^{\frac{m\lambda}{\alpha}} \left(\frac{\alpha}{m\lambda}\right)^{\frac{t}{\alpha}} \int_{\frac{m\lambda}{\alpha}}^{+\infty} v^{\frac{t}{\alpha}} e^{-v} dv$$
$$= e^{\frac{m\lambda}{\alpha}} \left(\frac{\alpha}{m\lambda}\right)^{\frac{t}{\alpha}} \Gamma\left(\frac{t}{\alpha} + 1, \frac{m\lambda}{\alpha}\right),$$

where $\Gamma(d, x)$ denotes the complementary incomplete gamma function defined by $\Gamma(d, x) = \int_x^{+\infty} t^{d-1} e^{-t} dt$. Therefore, we can write

$$M_X(t) = \sum_{m=0}^{+\infty} w_m e^{\frac{m\lambda}{\alpha}} \left(\frac{\alpha}{m\lambda}\right)^{\frac{t}{\alpha}} \Gamma\left(\frac{t}{\alpha} + 1, \frac{m\lambda}{\alpha}\right).$$

Alternatively, using the moments of X, one can write

$$M_X(t) = \sum_{r=0}^{+\infty} \frac{t^r}{r!} \mu'_r = \sum_{r=0}^{+\infty} \sum_{m=0}^{+\infty} \frac{t^r}{r!} w_m \frac{m\lambda}{\alpha^{r+1}} e^{\frac{m\lambda}{\alpha}} \int_1^{+\infty} (\ln u)^r e^{-\frac{m\lambda}{\alpha} u} du.$$

3.8. Incomplete Moments and Mean Deviations

The rth incomplete moment of X is defined by $m_r(t) = \mathbb{E}\left(X^r \mathbf{1}_{\{X \leq t\}}\right) = \int_{-\infty}^t x^r f(x) dx$. Using Equation (9), we can express $m_r(t)$ as

$$m_r(t) = \sum_{m=0}^{+\infty} w_m \int_{-\infty}^t x^r h_m(x) dx.$$

Doing successively the change of variables $u = e^{\alpha x}$, we obtain

$$\int_{-\infty}^t x^r h_m(x) dx = \frac{m\lambda}{\alpha^{r+1}} e^{\frac{m\lambda}{\alpha}} \int_1^{e^{\alpha t}} (\ln u)^r e^{-\frac{m\lambda}{\alpha} u} du.$$

The mean deviation of X about the mean is given by

$$\delta_1 = \mathbb{E}(|X - \mu'_1|) = 2\mu'_1 F(\mu'_1) - 2m_1(\mu'_1),$$

where $m_1(t)$ denote the first incomplete moment. The mean deviation of X about the median $M = Q(1/2)$ is given by

$$\delta_2 = \mathbb{E}(|X - M|) = \mu'_1 - 2m_1(M).$$

3.9. Entropies

Let us now investigate different kinds of entropies. For the sake of simplicity in exposition, we suppose that $c \in (0, 1)$ (the case $c > 1$ can be considered in a similar way). The Rényi entropy of X is defined by

$$\mathcal{I}_\gamma(X) = \frac{1}{1-\gamma} \ln\left[\int_{-\infty}^{+\infty} [f(x)]^\gamma dx\right],$$

with $\gamma > 0$ and $\gamma \neq 1$. It follows from (3) that

$$[f(x)]^\gamma = \frac{c^{a\gamma} [g(x)]^\gamma [G(x)]^{\gamma(a-1)} [1 - G(x)]^{\gamma(b-1)}}{B(a,b)^\gamma [1 - (1-c)G(x)]^{\gamma(a+b)}}.$$

The generalized binomial formula implies that

$$\frac{[G(x)]^{\gamma(a-1)}}{[1-(1-c)G(x)]^{\gamma(a+b)}} = \sum_{k=0}^{+\infty} \binom{-\gamma(a+b)}{k}(-1)^k(1-c)^k[G(x)]^{k+\gamma(a-1)}.$$

Similarly, we have

$$[G(x)]^{k+\gamma(a-1)} = \sum_{\ell=0}^{+\infty}\binom{k+\gamma(a-1)}{\ell}(-1)^\ell[1-G(x)]^\ell.$$

Therefore,

$$[f(x)]^\gamma =$$
$$\frac{c^{a\gamma}}{B(a,b)^\gamma} \sum_{k=0}^{+\infty}\sum_{\ell=0}^{+\infty}\binom{-\gamma(a+b)}{k}\binom{k+\gamma(a-1)}{\ell}(-1)^{k+\ell}(1-c)^k[1-G(x)]^{\ell+\gamma(b-1)}[g(x)]^\gamma.$$

By doing the change of variable $u = e^{\alpha x}$ and the change of variable $v = (\ell + \gamma b)\frac{\lambda}{\alpha}u$, we get

$$\int_{-\infty}^{+\infty} [1-G(x)]^{\ell+\gamma(b-1)}[g(x)]^\gamma dx = \int_0^{+\infty} e^{-(\ell+\gamma b)\frac{\lambda}{\alpha}(e^{\alpha x}-1)}\lambda^\gamma e^{\alpha\gamma x}dx$$

$$= \lambda^\gamma \frac{1}{\alpha}e^{(\ell+\gamma b)\frac{\lambda}{\alpha}} \int_1^{+\infty} u^{\gamma-1}e^{-(\ell+\gamma b)\frac{\lambda}{\alpha}u}du$$

$$= \frac{\alpha^{\gamma-1}}{(\ell+\gamma b)^\gamma}e^{(\ell+\gamma b)\frac{\lambda}{\alpha}} \int_{(\ell+\gamma b)\frac{\lambda}{\alpha}}^{+\infty} v^{\gamma-1}e^{-v}dv$$

$$= \frac{\alpha^{\gamma-1}}{(\ell+\gamma b)^\gamma}e^{(\ell+\gamma b)\frac{\lambda}{\alpha}}\Gamma\left(\gamma, (\ell+\gamma b)\frac{\lambda}{\alpha}\right).$$

By putting the above equalities together, we have

$$\mathcal{I}_\gamma(X) =$$
$$\frac{1}{1-\gamma}\left[\alpha\gamma\ln(c) - \gamma\ln(B(a,b)) + (\gamma-1)\ln(\alpha) + \frac{\gamma b\lambda}{\alpha}\right.$$
$$+ \ln\left.\left[\sum_{k=0}^{+\infty}\sum_{\ell=0}^{+\infty}\binom{-\gamma(a+b)}{k}\binom{k+\gamma(a-1)}{\ell}(-1)^{k+\ell}(1-c)^k\frac{e^{\ell\frac{\lambda}{\alpha}}}{(\ell+\gamma b)^\gamma}\Gamma\left(\gamma, (\ell+\gamma b)\frac{\lambda}{\alpha}\right)\right]\right].$$

The Shannon entropy of X is defined by $S(X) = \mathbb{E}(-\ln[f(X)])$. It can be obtained by the formula $S(X) = \lim_{\gamma\to 1^+} \mathcal{I}_\gamma(X)$.

The γ-entropy is defined by

$$H_\gamma(X) = \frac{1}{\gamma-1}\ln\left[1 - \int_{-\infty}^{+\infty}[f(x)]^\gamma dx\right].$$

Using the expansion above, we obtain

$$H_\gamma(X) = \frac{1}{\gamma-1}\ln\left[1 - \frac{c^{a\gamma}\alpha^{\gamma-1}e^{\gamma b\frac{\lambda}{\alpha}}}{B(a,b)^\gamma} \times\right.$$
$$\left.\sum_{k=0}^{+\infty}\sum_{\ell=0}^{+\infty}\binom{-\gamma(a+b)}{k}\binom{k+\gamma(a-1)}{\ell}(-1)^{k+\ell}(1-c)^k\frac{e^{\ell\frac{\lambda}{\alpha}}}{(\ell+\gamma b)^\gamma}\Gamma\left(\gamma, (\ell+\gamma b)\frac{\lambda}{\alpha}\right)\right].$$

3.10. Order Statistics

Let $X_1, \ldots, X_n$ be the random sample from X and $X_{i:n}$ be the ith order statistic. Then, the pdf of $X_{i:n}$ is given by

$$f_{i:n}(x) = \frac{n!}{(i-1)!(n-i)!} f(x)[F(x)]^{i-1}[1-F(x)]^{n-i}$$

$$= \frac{n!}{(i-1)!(n-i)!} \sum_{j=0}^{n-i} \binom{n-i}{j}(-1)^j f(x)[F(x)]^{j+i-1}.$$

It follows from Equations (8) and (9) that

$$f_{i:n}(x) = \frac{n!}{(i-1)!(n-i)!} \sum_{j=0}^{n-i} \binom{n-i}{j}(-1)^j \sum_{m=0}^{+\infty} w_m h_m(x) \left[\sum_{k=0}^{+\infty} v_k(1-H_k(x))\right]^{j+i-1}.$$

Using a result from [29], power series raised to a positive power as follows

$$\left(\sum_{k=0}^{+\infty} a_k x^k\right)^n = \sum_{k=0}^{+\infty} d_{n,k} x^k,$$

where the coefficients $(d_{n,k})_{k \in \mathbb{N}}$ are determined from the recurrence equation: $d_{n,0} = a_0^n$ and, for any $m \geq 1$, $d_{n,m} = (1/(ma_0)) \sum_{k=1}^{m}(k(n+1)-m)a_k d_{n,m-k}$. Therefore, noticing that $1 - H_k(x) = \left(e^{-\frac{\lambda}{\alpha}(e^{\alpha x}-1)}\right)^k$, we have

$$\left[\sum_{k=0}^{+\infty} v_k(1-H_k(x))\right]^{j+i-1} = \sum_{k=0}^{+\infty} d_{j+i-1,k}(1-H_k(x)),$$

where $d_{j+i-1,0} = v_0^{j+i-1}$ and, for any $m \geq 1$, $d_{j+i-1,m} = \frac{1}{mv_0} \sum_{k=1}^{m}(k(j+i)-m)v_k d_{j+i-1,m-k}$. By combining the equalities above, we obtain

$$f_{i:n}(x) = \frac{n!}{(i-1)!(n-i)!} \sum_{j=0}^{n-i} \binom{n-i}{j}(-1)^j \sum_{m=0}^{+\infty} \sum_{k=0}^{+\infty} w_m d_{j+i-1,k} h_m(x)(1-H_k(x)).$$

Finally, one can observe that $h_m(x)(1-H_k(x)) = m\lambda e^{\alpha x} e^{-\frac{(m+k)\lambda}{\alpha}(e^{\alpha x}-1)} = \frac{m}{m+k} u_{m+k}(x)$, where $u_{m+k}(x)$ denotes the pdf of the Gompertz distribution with parameters $(m+k)\lambda$ and α. Thus, the pdf of ith order statistic of the MBGz distribution can be expressed as a linear combination of Gompertz pdfs, i.e.,

$$f_{i:n}(x) = \frac{n!}{(i-1)!(n-i)!} \sum_{j=0}^{n-i} \binom{n-i}{j}(-1)^j \sum_{m=0}^{+\infty} \sum_{k=0}^{+\infty} w_m d_{j+i-1,k} \frac{m}{m+k} u_{m+k}(x).$$

Let r be a positive integer. Then, the rth ordinary moment of $X_{i:n}$ can be expressed as

$$\mathbb{E}(X_{i:n}^r) = \int_{-\infty}^{+\infty} x^r f_{i:n}(x) dx$$

$$= \frac{n!}{(i-1)!(n-i)!} \sum_{j=0}^{n-i} \binom{n-i}{j}(-1)^j \sum_{m=0}^{+\infty} \sum_{k=0}^{+\infty} w_m d_{j+i-1,k} \frac{m\lambda}{\alpha^{r+1}} e^{\frac{(m+k)\lambda}{\alpha}} \int_1^{+\infty} (\ln u)^r e^{-\frac{(m+k)\lambda}{\alpha} u} du.$$

4. Statistical Inference

4.1. Maximum Likelihood Estimation

We now investigate the estimation of the parameters of the MBGz distribution. Let $x_1, \ldots, x_n$ be n observed values from the MBGz distribution and $\xi = (\lambda, \alpha, a, b, c)$ be the vector of unknown parameters. The log likelihood function is given by

$$\ell(\xi) = an\ln(c) + n\ln(\lambda) + \alpha \sum_{i=1}^{n} x_i - \frac{\lambda b}{\alpha} \sum_{i=1}^{n} (e^{\alpha x_i} - 1) + (a-1) \sum_{i=1}^{n} \ln\left(1 - e^{-\frac{\lambda}{\alpha}(e^{\alpha x_i} - 1)}\right)$$
$$- n\ln(B(a,b)) - (a+b) \sum_{i=1}^{n} \ln\left[1 - (1-c)\left(1 - e^{-\frac{\lambda}{\alpha}(e^{\alpha x_i} - 1)}\right)\right].$$

The maximum likelihood estimators of the parameters are obtained by maximizing the log likelihood function. They can be determined by solving the non-linear equations: $\frac{\partial}{\partial \lambda}\ell(\xi) = 0$, $\frac{\partial}{\partial \alpha}\ell(\xi) = 0$, $\frac{\partial}{\partial a}\ell(\xi) = 0$, $\frac{\partial}{\partial b}\ell(\xi) = 0$, $\frac{\partial}{\partial c}\ell(\xi) = 0$ with

$$\frac{\partial \ell(\xi)}{\partial \lambda} = \frac{n}{\lambda} - \frac{b}{\alpha} \sum_{i=1}^{n} (e^{\alpha x_i} - 1) + (a-1) \sum_{i=1}^{n} \frac{\frac{1}{\alpha}(e^{\alpha x_i} - 1)e^{-\frac{\lambda}{\alpha}(e^{\alpha x_i} - 1)}}{1 - e^{-\frac{\lambda}{\alpha}(e^{\alpha x_i} - 1)}}$$
$$+ (a+b) \sum_{i=1}^{n} \frac{(1-c)\frac{1}{\alpha}(e^{\alpha x_i} - 1)e^{-\frac{\lambda}{\alpha}(e^{\alpha x_i} - 1)}}{1 - (1-c)\left(1 - e^{-\frac{\lambda}{\alpha}(e^{\alpha x_i} - 1)}\right)},$$

$$\frac{\partial \ell(\xi)}{\partial \alpha} = \sum_{i=1}^{n} x_i - \frac{\lambda b}{\alpha} \sum_{i=1}^{n} \left[x_i e^{\alpha x_i} - \frac{1}{\alpha}(e^{\alpha x_i} - 1)\right]$$
$$+ (a-1) \sum_{i=1}^{n} \frac{\frac{\lambda}{\alpha}e^{-\frac{\lambda}{\alpha}(e^{\alpha x_i} - 1)}\left[x_i e^{\alpha x_i} - \frac{1}{\alpha}(e^{\alpha x_i} - 1)\right]}{1 - e^{-\frac{\lambda}{\alpha}(e^{\alpha x_i} - 1)}}$$
$$+ (a+b) \sum_{i=1}^{n} \frac{(1-c)e^{-\frac{\lambda}{\alpha}(e^{\alpha x_i} - 1)}\left[x_i e^{\alpha x_i} - \frac{1}{\alpha}(e^{\alpha x_i} - 1)\right]}{1 - (1-c)\left(1 - e^{-\frac{\lambda}{\alpha}(e^{\alpha x_i} - 1)}\right)},$$

by setting $B^{(1,0)}(a,b) = \frac{\partial}{\partial a} B(a,b)$ and $B^{(0,1)}(a,b) = \frac{\partial}{\partial b} B(a,b)$ (one can remark that $B^{(1,0)}(a,b) = \psi(a) - \psi(a+b)$ and $B^{(0,1)}(a,b) = \psi(b) - \psi(a+b)$, where $\psi(x)$ denotes the so called digamma function),

$$\frac{\partial \ell(\xi)}{\partial a} = n\ln c + \sum_{i=1}^{n} \ln\left(1 - e^{-\frac{\lambda}{\alpha}(e^{\alpha x_i} - 1)}\right) - n\frac{B^{(1,0)}(a,b)}{B(a,b)}$$
$$- \sum_{i=1}^{n} \ln\left[1 - (1-c)\left(1 - e^{-\frac{\lambda}{\alpha}(e^{\alpha x_i} - 1)}\right)\right],$$

$$\frac{\partial \ell(\xi)}{\partial b} = -\frac{\lambda}{\alpha} \sum_{i=1}^{n} (e^{\alpha x_i} - 1) - n\frac{B^{(0,1)}(a,b)}{B(a,b)} - \sum_{i=1}^{n} \ln\left[1 - (1-c)\left(1 - e^{-\frac{\lambda}{\alpha}(e^{\alpha x_i} - 1)}\right)\right]$$

and

$$\frac{\partial \ell(\xi)}{\partial c} = \frac{an}{c} - (a+b) \sum_{i=1}^{n} \frac{1 - e^{-\frac{\lambda}{\alpha}(e^{\alpha x_i} - 1)}}{1 + (1-c)\left(1 - e^{-\frac{\lambda}{\alpha}(e^{\alpha x_i} - 1)}\right)}.$$

We can solve the above non-linear equations simultaneously. A mathematical package can be used to get the maximum likelihood estimators of the unknown parameters. In addition, all the second-order derivatives exist. As usual, the asymptotic normality of the maximum likelihood estimators can be used to construct informative objects (approximate confidence intervals, confidence regions, and testing hypotheses of λ, α, a, b, c, etc.).

4.2. Simulation

From a theoretical point of view, the performances of the different estimates (MLEs) for the MBGz distribution are difficult to compare. We therefore propose a simulation study that uses their mean square errors (MSEs) for different sample sizes as benchmarks. The software package Mathematica (version 9) was used. Different sample sizes were considered through the experiments at size $n = 50$, 100 and 150. The experiment was repeated 3000 times. In each experiment, the estimates of the parameters were obtained by maximum likelihood methods of estimation. The means and MSEs for the different estimators can be found in Table 1. We observed that MSEs are decreasing with increasing n.

Table 1. The MLEs and MSEs of MBGz distribution.

n	Parameters	Initial	MLE	MSE	Initial	MLE	MSE
50	a	3.0	3.0024	0.5057	2.5	2.6424	0.1737
	b	1.5	1.6409	0.1499	1.5	1.5219	0.0400
	c	0.5	0.4941	0.0008	0.5	0.5050	0.0004
	α	0.5	0.5422	0.0198	0.5	0.5291	0.0116
	λ	0.5	0.5241	0.0387	0.5	0.5235	0.0122
100	a	3.0	3.0778	0.2458	2.5	2.5060	0.0754
	b	1.5	1.6083	0.0779	1.5	1.5373	0.0291
	c	0.5	0.4986	0.0003	0.5	0.4991	0.0003
	α	0.5	0.5572	0.0147	0.5	0.5123	0.0029
	λ	0.5	0.4926	0.0126	0.5	0.5035	0.0070
150	a	3.0	2.9041	0.1015	2.5	2.5125	0.0284
	b	1.5	1.6159	0.0485	1.5	1.5232	0.0088
	c	0.5	0.4940	0.0002	0.5	0.5002	0.0001
	α	0.5	0.5477	0.0094	0.5	0.5137	0.0015
	λ	0.5	0.4694	0.0072	0.5	0.4968	0.0015
50	a	1.5	1.4706	0.0325	1.5	1.5435	0.0641
	b	1.8	1.7764	0.0639	1.8	1.7838	0.0955
	c	0.5	0.5054	0.0013	1.5	1.5285	0.0203
	α	0.5	0.4833	0.0029	0.5	0.4895	0.0008
	λ	0.5	0.5488	0.0160	0.5	0.5364	0.0118
100	a	1.5	1.5138	0.0201	1.5	1.5194	0.0224
	b	1.8	1.8177	0.0380	1.8	1.8309	0.0451
	c	0.5	0.5004	0.0007	1.5	1.5010	0.0047
	α	0.5	0.5007	0.0023	0.5	0.5011	0.0005
	λ	0.5	0.5106	0.0059	0.5	0.5036	0.0028
150	a	1.5	1.5313	0.0102	1.5	1.4690	0.0094
	b	1.8	1.8152	0.0194	1.8	1.8396	0.0258
	c	0.5	0.5055	0.0003	1.5	1.4864	0.0017
	α	0.5	0.5173	0.0022	0.5	0.5007	0.0004
	λ	0.5	0.5044	0.0034	0.5	0.4943	0.0009

4.3. Applications

This section provides an application to show how the MBGz distribution can be applied in practice. We compared MBGz to Exponentaited Generalized Weibull–Gompertz distribution (EGWGz) by El-Bassiouny et al. [30] and other well known distributions in literature, Kumaraswamy–Gompertz (Kw-Gz), beta Gompertz (BGz) and Gompertz (Gz) models. The MLEs are computed using Quasi-Newton Code for Bound Constrained Optimization and the log-likelihood function evaluated. The goodness-of-fit measures, Anderson–Darling (A*), Cramer–von Mises (W*), Akaike Information Criterion (AIC), Bayesian Information Criterion (BIC), and log-likelihood ($\hat{\ell}$) values are computed. As usual, the lower are the values of these criteria, the better is the fit. In addition, the value for the Kolmogorov–Smirnov (KS) statistic and its *p*-value are reported. The required computations are carried out in the R software (version 3).

4.3.1. Dataset 1

The first dataset was given in [31]. It represents the time to failure of turbocharger of a certain type of engine. The dataset is as follows: 0.0312, 0.314, 0.479, 0.552, 0.700, 0.803, 0.861, 0.865, 0.944, 0.958, 0.966, 0.977, 1.006, 1.021, 1.027, 1.055, 1.063, 1.098, 1.140, 1.179, 1.224, 1.240, 1.253, 1.270, 1.272, 1.274, 1.301, 1.301, 1.359, 1.382, 1.382, 1.426, 1.434, 1.435, 1.478, 1.490, 1.511, 1.514, 1.535, 1.554, 1.566, 1.570, 1.586, 1.629, 1.633, 1.642, 1.648, 1.684, 1.697, 1.726, 1.770, 1.773, 1.800, 1.809, 1.818, 1.821, 1.848, 1.880, 1.954, 2.012, 2.067, 2.084, 2.090, 2.096, 2.128, 2.233, 2.433, 2.585, 2.585.

4.3.2. Dataset 2

The second dataset was considered in [32]. It corresponds to a single fiber with 20 and 101 mm of gauge length, respectively. The dataset is as follows: 1.6, 2.0, 2.6, 3.0, 3.5, 3.9, 4.5, 4.6, 4.8, 5.0, 5.1, 5.3, 5.4, 5.6, 5.8, 6.0, 6.0, 6.1, 6.3, 6.5, 6.5, 6.7, 7.0, 7.1, 7.3, 7.3, 7.3, 7.7, 7.7, 7.8, 7.9, 8.0, 8.1, 8.3, 8.4, 8.4, 8.5, 8.7, 8.8, 9.0.

Tables 2 and 3 present the maximum likelihood estimates, with the corresponding standard errors in parentheses, of the unknown parameters (λ, α, a, b, c) of the MBGz distribution for Datasets 1 and 2, respectively. Tables 4 and 5 show the statistics AIC, BIC, W^*, A^*, KS, and *p*-Value values for all the considered models. We then see that the proposed MBGz model fits the considered data better than the other models. Thus, the proposed MBGz model provides an interesting alternative to other existing models for modeling positive real data. To complete this fact, PP, QQ, epdf and ecdf plots of the MBGz distribution are given in Figures 2 and 3 for Datasets 1 and 2, respectively.

Table 2. MLEs (standard errors in parentheses) for Dataset 1.

Distribution	Estimates				
MBGz (λ, α, a, b, c)	0.0085	2.5537	1.0737	1.3153	5.0687
	(0.0067)	(0.5727)	(0.3197)	(0.8933)	(3.3003)
EGWGz (λ, a, b, c, β)	3.2078	2.4598	0.0203	1.8974	0.5460
	(1.2099)	(0.6498)	(0.0531)	(1.8193)	(0.2430)
KwGz (a, b, c, d, θ)	0.1861	1.4948	1.4909	0.9811	
	(0.3130)	(0.5076)	(0.4735)	(2.4368)	
BGz (a, b, θ, v)	0.3144	1.5591	1.4798	0.4966	
	(0.4283)	(0.3658)	(0.4543)	(0.8692)	
Gz (λ, α)	0.0841	1.8811			
	(0.0268)	(0.2043)			

Table 3. MLEs (standard errors in parentheses) for Dataset 2.

Distribution	Estimates				
MBGz (λ, α, a, b, c)	0.0098	0.5270	0.8768	4.5635	0.1561
	(0.0116)	(0.1599)	(0.3893)	(0.8862)	(0.2442)
EGWGz (λ, a, b, c, β)	0.0101	0.6077	0.1078	1.6929	0.6613
	(0.0141)	(0.1506)	(0.3427)	(1.2539)	(0.3379)
KwGz (a, b, θ, v)	0.0133	0.2923	2.0164	13.7085	
	(0.0120)	(0.1641)	(0.7880)	(7.0208)	
BGz (a, b, c, d, θ)	0.0125	0.1856	3.7622	2.0116	
	(0.0100)	(0.1601)	(2.5635)	(3.3802)	
Gz (λ, α)	0.0074	(0.6243)			
	(0.0035)	(0.0748)			

Table 4. The AIC, BIC, W^*, A^*, KS, and p-Value values for Dataset 1.

Dist	ℓ	AIC	BIC	W^*	A^*	KS	p-Value
MBGz	50.0387	110.0776	118.2481	0.0328	0.2745	0.0539	0.9889
EGWGz	52.6888	115.3776	126.5482	0.0706	0.5341	0.0785	0.7885
KwGz	51.2042	110.4084	119.3448	0.0529	0.4125	0.0640	0.9396
BGz	51.1518	110.3026	119.2399	0.0518	0.4057	0.0627	0.9484
Gz	53.9686	111.9374	122.4056	0.0819	0.5921	0.0810	0.7547

Table 5. The AIC, BIC, W^*, A^*, KS, and p-Value values for Dataset 2.

Dist	ℓ	AIC	BIC	W^*	A^*	KS	p-Value
MBGz	78.2184	168.1770	176.0214	0.0222	0.1840	0.0707	0.9888
EGWGz	79.3744	168.5489	178.5933	0.0479	0.2922	0.0821	0.9623
KwGz	80.7197	169.4395	176.1950	0.0430	0.3326	0.0966	0.8489
BGz	82.9924	173.9849	180.7404	0.0922	0.6736	0.1080	0.7389
Gz	80.9566	168.9234	177.2911	0.0359	0.2335	0.0903	0.8299

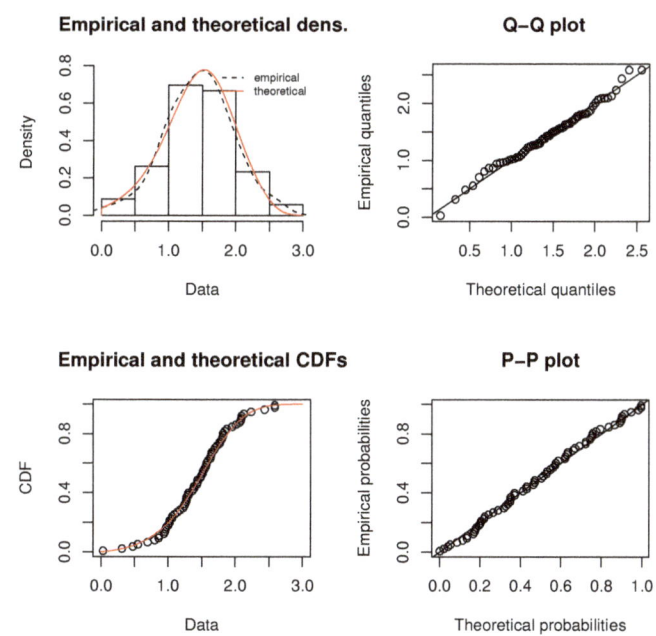

Figure 2. PP, QQ, epdf and ecdf plots of the MBGz distribution for Dataset 1.

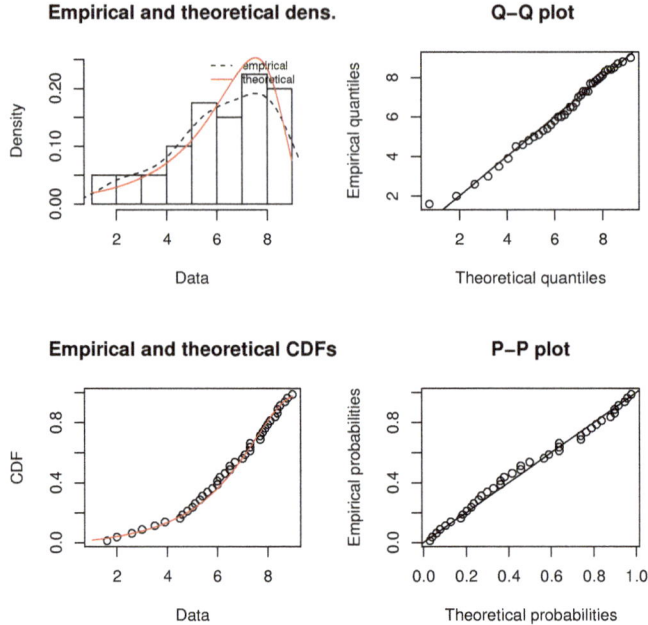

Figure 3. PP, QQ, epdf and ecdf plots of the MBGz distribution for Dataset 2.

Author Contributions: I.E., F.J., C.C., M.E. and S.A. have contributed equally to this work.

Funding: This research received no external funding.

Acknowledgments: The authors would like to thank the referees for their valuable comments which helped to improve the manuscript.

Conflicts of Interest: The authors declare no conflict of interest.

References

1. Gompertz, B. On the nature of the function expressive of the law of human mortality and on the new mode of determining the value of life contingencies. *Philos. Trans. R. Soc. A* **1825**, *115*, 513–580. [CrossRef]
2. Tjørve, K.M.C.; Tjørve, E. The use of Gompertz models in growth analyses, and new Gompertz-model approach: An addition to the Unified-Richards family. *PLoS ONE* **2017**, *12*. [CrossRef]
3. El-Gohary, A.; Alshamrani, A.; Al-Otaibi, A. The generalized Gompertz distribution. *Appl. Math. Model.* **2013**, *37*, 13–24. [CrossRef]
4. Jafari, A.A.; Tahmasebi, S.; Alizadeh, M. The beta-Gompertz distribution. *Revista Colombiana de Estadistica* **2014**, *37*, 141–158. [CrossRef]
5. Eugene, N.; Lee, C.; Famoye, F. Beta-normal distribution and its applications. *Comm. Statist. Theory Methods* **2002**, *31*, 497–512. [CrossRef]
6. Benkhelifa, L. The beta generalized Gompertz distribution. *Appl. Math. Model.* **2017**, *52*, 341–357. [CrossRef]
7. El-Damcese, M.A.; Mustafa, A.; El-Desouky, B.S.; Mustafa, M.E. Generalized Exponential Gompertz Distribution. *Appl. Math.* **2015**, *6*, 2340–2353. [CrossRef]
8. Tahir, M.H.; Cordeiro, G.M.; Alizadeh, M.; Mansoor, M.; Zubair, M.; Hamedani, G.G. The Odd Generalized Exponential Family of Distributions with Applications. *J. Stat. Distrib. Appl.* **2015**, *2*, 1–28. [CrossRef]
9. Roozegar, R.; Tahmasebi, S.; Jafari, A.A. The McDonald Gompertz distribution: Properties and applications. *Commun. Stat. Simul. Comput.* **2017**, *46*, 3341–3355. [CrossRef]

10. Alexander, C.; Cordeiro, G.M.; Ortega, E.M.M.; Sarabia, J.M. Generalized beta-generated distributions. *Comput. Stat. Data Anal.* **2012**, *56*, 1880–1897. [CrossRef]
11. Khan, M.S.; King, R.; Hudson, I.L. Transmuted Generalized Gompertz distribution with application. *J. Stat. Theory Appl.* **2017**, *16*, 65–80. [CrossRef]
12. Moniem, A.I.B.; Seham, M. Transmuted Gompertz Distribution. *Comput. Appl. Math.* **2015**, *1*, 88–96.
13. Shaw, W.; Buckley, I. *The Alchemy of Probability Distributions: Beyond Gram-Charlier Expansions and a Skewkurtotic-Normal Distribution from a Rank Transmutation Map*; Research Report; King's College: London, UK, 2007.
14. Chukwu, A.U.; Ogunde, A.A. On kumaraswamy gompertz makeham distribution. *Am. J. Math. Stat.* **2016**, *6*, 122–127.
15. Lima, F.P.; Sanchez, J.D.; da Silva, R.C.; Cordeiro, G.M. The Kumaraswamy Gompertz distribution. *J. Data Sci.* **2015**, *13*, 241–260.
16. Benkhelifa, L. Marshall-Olkin extended generalized Gompertz distribution. *J. Data Sci.* **2017**, *15*, 239–266.
17. Yaghoobzadeh, S. A new generalization of the Marshall-Olkin Gompertz distribution. *Int. J. Syst. Assur. Eng. Manag.* **2017**, *8*, 1580–1587. [CrossRef]
18. Marshall, A.W.; Olkin, I. A new method for adding a parameter to a family of distributions with application to the exponential and Weibull families. *Biometrika* **1997**, *84*, 641–652. [CrossRef]
19. Shadrokh, A.; Yaghoobzadeh, S.S. The Beta Gompertz Geometric Distribution: Mathematical Properties and Applications. *Andishe_ye Amari* **2018**, *22*, 81–91.
20. Nadarajah, S.; Teimouri, M.; Shih, S.H. Modified Beta Distributions. *Sankhya Ser. B* **2014**, *76*, 19–48. [CrossRef]
21. Alizadeh, M.; Cordeiro, G.M.; Brito, E. The beta Marshall-Olkin family of distributions. *J. Stat. Distrib. Appl.* **2015**, *2*, 4. [CrossRef]
22. Nadarajah, S.; Kotz, S. The beta exponential distribution. *Reliab. Eng. Syst. Saf.* **2006**, *91*, 689–697. [CrossRef]
23. Gupta, R.D.; Kundu, D. Generalized Exponential Distributions. *Aust. N. Z. J. Stat.* **1999**, *41*, 173–188. [CrossRef]
24. Kenney, J.F.; Keeping, E.S. *Mathematics of Statistics*, 3rd ed.; Chapman and Hall Ltd.: Rahway, NJ, USA, 1962.
25. Moors, J.J. A quantile alternative for kurtosis. *J. R. Stat. Soc. Ser. D* **1988**, *37*, 25–32. [CrossRef]
26. Milgram, M. The generalized integro-exponential function. *Math. Comput.* **1985**, *44*, 443–458. [CrossRef]
27. Lenart, A. The moments of the Gompertz distribution and maximum likelihood estimation of its parameters. *Scand. Actuar. J.* **2014**, *3*, 255–277. [CrossRef]
28. MacGillivray, H.L. Skewness and Asymmetry: Measures and Orderings. *Ann. Stat.* **1986**, *14*, 994–1011. [CrossRef]
29. Gradshteyn, I.S.; Ryzhik, I.M. *Table of Integrals, Series and Products*; Academic Press: New York, NY, USA, 2000.
30. El-Bassiouny, H.; EL-Damcese, M.; Mustafa, A.; Eliwa, M.S. Exponentiated Generalized Weibull-Gompertz Distribution with Application in Survival Analysis. *J. Stat. Appl. Probab.* **2017**, *6*, 7–16. [CrossRef]
31. Xu, K.; Xie, M.; Tang, L.C.; Ho, S.L. Application of Neural Networks in forecasting Engine Systems Reliability. *Appl. Soft Comput.* **2003**, *2*, 255–268. [CrossRef]
32. Badar, M.G.; Priest, A.M. Statistical aspects of fiber and bundle strength in hybrid composites. In *Progress in Science and Engineering Composites, Proceedings of the ICCM-IV, Tokyo, Japan, 25–28 October 1982*; Hayashi, T., Kawata, K., Umekawa, S., Eds.; Japan Society for Composite Materials: Tokyo, Japan, 1982; pp. 1129–1136.

 © 2018 by the authors. Licensee MDPI, Basel, Switzerland. This article is an open access article distributed under the terms and conditions of the Creative Commons Attribution (CC BY) license (http://creativecommons.org/licenses/by/4.0/).

Article

Computation of Probability Associated with Anderson–Darling Statistic

Lorentz Jäntschi [1,2] **and Sorana D. Bolboacă** [3,*]

1 Department of Physics and Chemistry, Technical University of Cluj-Napoca, Muncii Blvd. No. 103-105, Cluj-Napoca 400641, Romania; lorentz.jantschi@gmail.com
2 Doctoral Studies, Babeş-Bolyai University, Mihail Kogălniceanu Str., No. 1, Cluj-Napoca 400028, Romania
3 Department of Medical Informatics and Biostatistics, Iuliu Haţieganu University of Medicine and Pharmacy, Louis Pasteur Str., No. 6, Cluj-Napoca 400349, Romania
* Correspondence: sbolboaca@umfcluj.ro; Tel.: +40-766-341-408

Received: 14 April 2018; Accepted: 23 May 2018; Published: 25 May 2018

Abstract: The correct application of a statistical test is directly connected with information related to the distribution of data. Anderson–Darling is one alternative used to test if the distribution of experimental data follows a theoretical distribution. The conclusion of the Anderson–Darling test is usually drawn by comparing the obtained statistic with the available critical value, which did not give any weight to the same size. This study aimed to provide a formula for calculation of *p*-value associated with the Anderson–Darling statistic considering the size of the sample. A Monte Carlo simulation study was conducted for sample sizes starting from 2 to 61, and based on the obtained results, a formula able to give reliable probabilities associated to the Anderson–Darling statistic is reported.

Keywords: Anderson–Darling test (AD); probability; Monte Carlo simulation

1. Introduction

Application of any statistical test is made under certain assumptions, and violation of these assumptions could lead to misleading interpretations and unreliable results [1,2]. One main assumption that several statistical tests have is related with the distribution of experimental or observed data (H_0 (null hypothesis): The data follow the specified distribution vs. H_1 (alternative hypothesis): The data do not follow the specified distribution). Different tests, generally called "goodness-of-fit", are used to assess whether a sample of observations can be considered as a sample from a given distribution. The most frequently used goodness-of-fit tests are Kolmogorov–Smirnov [3,4], Anderson–Darling [5,6], Pearson's chi-square [7], Cramér–von Mises [8,9], Shapiro–Wilk [10], Jarque–Bera [11–13], D'Agostino–Pearson [14], and Lilliefors [15,16]. The goodness-of-fit tests use different procedures (see Table 1). Alongside the well-known goodness-of-fit test, other methods based for example on entropy estimator [17–19], jackknife empirical likelihood [20], on the prediction of residuals [21], or for testing multilevel survival data [22] or multilevel models with binary outcomes [23] have been reported in the scientific literature.

Table 1. The goodness-of-fit tests: approaches.

Test Name	Abbreviation	Procedure
Kolmogorov–Smirnov	KS	Proximity analysis of the empirical distribution function (obtained on the sample) and the hypothesized distribution (theoretical)
Anderson–Darling	AD	How close the points are to the straight line estimated in a probability graphic
chi-square	CS	Comparison of sample data distribution with a theoretical distribution
Cramér–von Mises	CM	Estimation of the minimum distance between theoretical and sample probability distribution
Shapiro–Wilk	SW	Based on a linear model between the ordered observations and the expected values of the ordered statistics of the standard normal distribution
Jarque–Bera	JB	Estimation of the difference between asymmetry and kurtosis of observed data and theoretical distribution
D'Agostino–Pearson	AP	Combination of asymmetry and kurtosis measures
Lilliefors	LF	A modified KS that uses a Monte Carlo technique to calculate an approximation of the sampling distribution

Tests used to assess the distribution of a dataset received attention from many researchers (for testing normal or other distributions) [24–27]. The normal distribution is of higher importance, since the resulting information will lead the statistical analysis on the pathway of parametric or non-parametric tests [28–33]. Different normality tests are implemented on various statistical packages (e.g., Minitab—http://www.minitab.com/en-us/; EasyFit—http://www.mathwave.com/easyfit-distribution-fitting.html; Develve—http://develve.net/; r("nortest" nortest)—https://cran.r-project.org/web/packages/nortest/nortest.pdf; etc.).

Several studies aimed to compare the performances of goodness-of-fit tests. In a Monte Carlo simulation study conducted on the normal distribution, Kolmogorov–Smirnov test has been identified as the least powerful test, while opposite Shapiro–Wilks test was identified as the most powerful test [34]. Furthermore, Anderson–Darling test was found to be the best option among five normality tests whenever t-statistics were used [35]. More weight to the tails are given by the Anderson–Darling test compared to Kolmogorov–Smirnov test [36]. The comparisons between different goodness-of-fit tests is frequently conducted by comparing their power [37,38], using or not confidence intervals [39], distribution of p-values [40], or ROC (receiver operating characteristic) analysis [32].

The interpretation of the Anderson–Darling test is frequently made by comparing the AD statistic with the critical value for a particular significance level (e.g., 20%, 10%, 5%, 2.5%, or 1%) even if it is known that the critical values depend on the sample size [41,42]. The main problem with this approach is that the critical values are available just for several distributions (e.g., normal and Weibull distribution in Table 2 [43], generalized extreme value and generalized logistic [44], etc.) but could be obtained in Monte Carlo simulations [45]. The primary advantage of the Anderson–Darling test is its applicability to test the departure of the experimental data from different theoretical distributions, which is the reason why we decided to identify the method able to calculate its associated p-value as a function also of the sample size.

D'Augostino and Stephens provided different formulas for calculation of p-values associated to the Anderson–Darling statistic (AD), along with a correction for small sample size (AD*) [37]. Their equations are independent of the tested theoretical distribution and highlight the importance of the sample size (Table 3).

Several Excel implementations of Anderson–Darling statistic are freely available to assist the researcher in testing if data follow, or do not follow, the normal distribution [46–48]. Since almost all distributions are dependent by at least two parameters, it is not expected that one goodness-of-fit test will provide sufficient information regarding the risk of error, because using only one method (one test) gives the expression of only one constraint between parameters. In this regard, the example provided in [49] is illustrative, and shows how the presence of a single outlier induces complete disarray between statistics, and even its removal does not bring the same risk of error as a result of applying different goodness-of-fit tests. Given this fact, calculation of the combined probability of independent (e.g., independent of the tested distribution) goodness-of-fit tests [50,51] is justified.

Good statistical practice guidelines request reporting the p-value associated with the statistics of a test. The sample size influences the p-value of statistics, so its reporting is mandatory to assure a proper interpretation of the statistical results. Our study aimed to identify, assess, and implement an explicit function of the p-value associated with the Anderson–Darling statistic able to take into consideration both the value of the statistic and the sample size.

Table 2. Anderson–Darling test: critical values according to significance level.

Distribution [Ref]	$\alpha = 0.10$	$\alpha = 0.05$	$\alpha = 0.01$
Normal & lognormal [43]	0.631	0.752	1.035
Weibull [43]	0.637	0.757	1.038
Generalized extreme value [44]	-	-	-
$n = 10$	0.236	0.276	0.370
$n = 20$	0.232	0.274	0.375
$n = 30$	0.232	0.276	0.379
$n = 40$	0.233	0.277	0.381
$n = 50$	0.233	0.277	0.383
$n = 100$	0.234	0.279	0.387
Generalized logistic [44]	-	-	-
$n = 10$	0.223	0.266	0.374
$n = 20$	0.241	0.290	0.413
$n = 30$	0.220	0.301	0.429
$n = 40$	0.254	0.306	0.435
$n = 50$	0.258	0.311	0.442
$n = 100$	0.267	0.323	0.461
Uniform [52] *	1.936	2.499	3.903

* Expressed as upper tail percentiles.

Table 3. Anderson–Darling for small sizes: p-values formulas.

Anderson–Darling Statistic	Formula for p-Value Calculation
$AD \geq 0.6$	$\exp(1.2937 - 5.709 \cdot (AD^*) + 0.0186 \cdot (AD^*)^2)$
$0.34 < AD^* < 0.6$	$\exp(0.9177 - 4.279 \cdot (AD^*) - 1.38 \cdot (AD^*)^2)$
$0.2 < AD^* < 0.34$	$1 - \exp(-8.318 + 42.796 \cdot (AD^*) - 59.938 \cdot (AD^*)^2)$
$AD^* \leq 0.2$	$1 - \exp(-13.436 + 101.14 \cdot (AD^*) - 223.73 \cdot (AD^*)^2)$

$AD* = AD(1 + \frac{0.75}{n} + \frac{2.25}{n^2})$; $AD = -n - \frac{1}{n} \sum_{i=0}^{n}(2 \cdot i - 1) \cdot [\ln(F(X_i)) + \ln(1 - F(X_{n-i+1}))]$.

2. Materials and Methods

2.1. Anderson–Darling Order Statistic

For a sample $Y = (y_1, y_2, \ldots, y_n)$, the data are sorted in ascending order (let $X = \text{Sort}(Y)$, and then $X = (x_1, x_2, \ldots, x_n)$ with $x_i \leq x_{i+1}$ for $0 < i < n$, and $x_i = y_{\sigma(i)}$, where σ is a permutation of $\{1, 2, \ldots, n\}$ which makes the X series sorted). Let the CDF be the associated cumulative distribution function and InvCDF the inverse of this function for any PDF (probability density function). The series $P = (p_1, p_2, \ldots, p_n)$ defined by $p_i = \text{InvCDF}(x_i)$ (or $Q = (q_1, q_2, \ldots, q_n)$ defined by $q_i = \text{InvCDF}(y_i)$, where the P is the unsorted array, and Q is the sorted array) are samples drawn from a uniform distribution only if Y (and X) are samples from the distribution with PDF.

At this point, the order statistics are used to test the uniformity of P (or for Q), and for this reason, the values of X are ordered (in Y). On the ordered probabilities (on P), several statistics can be computed, and Anderson–Darling (AD) is one of them:

$$AD = AD(P, n) = -n - \sum_{i=1}^{n} \frac{(2i-1)\ln(p_i(1-p_{n-i+1}))}{n}. \tag{1}$$

The associated AD statistic for a "perfect" uniform distribution can be computed after splitting the [0, 1] interval into n equidistant intervals (i/n, with $0 \leq i \leq n$ being their boundaries) and using the middles of those intervals $r_i = (2i-1)/2n$:

$$AD_{\min}(n) = AD(R, n) = -n + 4H_1(R, n). \tag{2}$$

where H_1 is the Shannon entropy for R in nats (the units of information or entropy) ($H_1(R,n) = -\Sigma r_i \cdot \ln(r_i)$).

Equation (2) gives the smallest possible value for AD. The value of the AD increases with the increase of the departure between the perfect uniform distribution and the observed distribution (P).

2.2. Monte Carlo Experiment for Anderson–Darling Statistic

The probability associated with a particular value of the AD statistic can be obtained using a Monte Carlo experiment. The AD statistics are calculated for a large enough number of samples (let be m the number of samples), the values are sorted, and then the relative position of the observed value of the AD in the series of Monte Carlo-calculated values gives the probability associated with the statistic of the AD test.

It should be noted that the equation linking the statistic and the probability also contains the size of the sample, and therefore, the probability associated with the AD value is dependent on n.

Taking into account all the knowledge gains until this point, it is relatively simple to do a Monte Carlo experiment for any order statistic. The only remaining problem is how to draw a sample from a uniform distribution in such way as to not affect the outcome. One alternative is to use a good random generator, such as Mersenne Twister [53], and this method was used to generate our samples as an alternative to the stratified random approach.

2.3. Stratified Random Strategy

Let us assume that three numbers (t_1, t_2, t_3) are extracted from a [0, 1) interval using Mersenne Twister method. Each of those numbers can be <0.5 or ≥0.5, providing 2^3 possible cases (Table 4).

Table 4. Cases for the half-split of [0, 1).

Class	t_1	t_2	t_3	Case
	0	0	0	1
	0	0	1	2
	0	1	0	3
"0" if $t_i < 0.5$	0	1	1	4
"1" if $t_i \geq 0.5$	1	0	0	5
	1	0	1	6
	1	1	0	7
	1	1	1	8

It is not a good idea to use the design presented in Table 4 in its crude form, since it is transformed to a problem with an exponential (2^n) complexity. The trick is to observe the pattern in Table 4. In fact, for (n + 1) cases, with different frequencies of occurrence following the model, the results are given in Table 5.

Table 5. Unique cases for the half-split of [0, 1).

$|\{t_i \mid t_i < 0.5\}|$	$|\{t_i \mid t_i \geq 0.5\}|$	Frequency (Case in Table 4)
3	0	1 (case 1)
2	1	3 (case 2, 3, 5)
1	2	3 (case 4, 6, 7)
0	3	1 (case 8)

The complexity of the problem of enumerating all the cases stays with the design presented in Table 5 at the same order of magnitude with n (we need to list only n + 1 cases instead of 2^n).

The frequencies listed in Table 5 are combinations of n objects taken by two (intervals), so instead of enumerating all 2^n cases, it is enough to record only n + 1 cases weighted with their relative occurrence.

The effect of the pseudo-random generator is significantly decreased (the decrease is a precise order of magnitude of the binary representation, one unit in $\log_2$ transformation: $1 = \log_2 2$, for the (0, 0.5) and (0.5, 1) split) by doing a stratified random sample.

The extractions of a number from (0, 0.5) and from (0.5, 1) were furthermore made in our experiment with Mersenne Twister random (if x = Random() with $0 \leq x < 1$ then $0 \leq x/2 < 1$ and $0.5 \leq 0.5 + x/2 < 1$). Table 5 provides all the information we need to do the design. For any n, for k from 0 to n, exactly k numbers are generated as Random()/2, and sorted. Furthermore, exactly $n-k$ numbers are generated as 0.5 + Random()/2, and the frequency associated with this pattern is $n!/(k! \cdot (n-k)!)$.

The combinations can also be calculated iteratively: cnk(n,0) = 1, and cnk(n,k) = cnk(n,(k − 1))·(n − k + 1)/k for successive $1 \leq k \leq n$.

2.4. Model for Anderson–Darling Statistic

Performing the Monte Carlo (MC) experiment (generates, analyzes, and provides the outcome) each time when a probability associated with the AD statistic is needed is resource-consuming and not effective. For example, if we generate for a certain sample size (n) a large number of samples $m = 1.28 \times 10^{10}$, then the needed storage space is 51.2 Gb for each n. Given 1 Tb of storage capacity, it can store only 20 iterations of n, as in the series of the AD(n). However, this is not needed, since it is possible to generate and store the results of the Monte Carlo analysis, but a proper model is required.

It is not necessary to have a model for any probability, since the standard thresholds for rejecting an agreement are commonly set to α = 0.2, 0.1, 0.05, 0.02, 0.01 ($\alpha = 1 - p$). A reliable result could be considered the model for the AD when $p \geq 0.5$. Therefore, the AD (as AD = AD(n,p)) for 501 value of the p from 0.500 to 0.001, and for n from 2 to 61 were extracted, tabulated, and used to develop the model.

A search for a dependency of AD = AD(p) (or p = p(AD)) for a particular n may not reveal any pattern. However, if the value of the statistic is exponentiated (see the ln function in the AD formula), values for the model start to appear (see Figure 1a) after a proper transformation of p. On the other hand, for a given n, an inconvenience for the AD(p) (or for its inverse, p = p(AD)) is to have on the plot, a non-uniform repartition of the points—for instance, precisely two points for $5 \leq$ AD < 6 and 144 points for AD < 1. As a consequence, any method trying to find the best fit based on this raw data will fail because it will give too much weight on the lower part with a much higher concentration of the points. The problem is the same for exp(AD) replacing AD (Figure 1b) but is no more the case for $1/(1 - p)$ as a function of exp(AD) (Figure 1c), since the dependence begins to look like a linear one. Figure 1b suggests that a logarithm on both axes will reduce the difference in the concentration of points in the intervals (Figure 1d), but at this point, is not necessary to apply it, since the last spots in Figure 1c may act as "outliers" trailing the slope. A good fit in the rarefied region of high p (and low α) is desired. It is not so important if we will have a 1% error at p = 50%, but is essential not to have a 1% error at p = 99% (the error will be higher than the estimated probability, $\alpha = 1 - p$. Therefore, in this case (Figure 1c), big numbers (e.g., ~200, 400) will have high values of residuals, and will trail the model to fit better in the rarefied region.

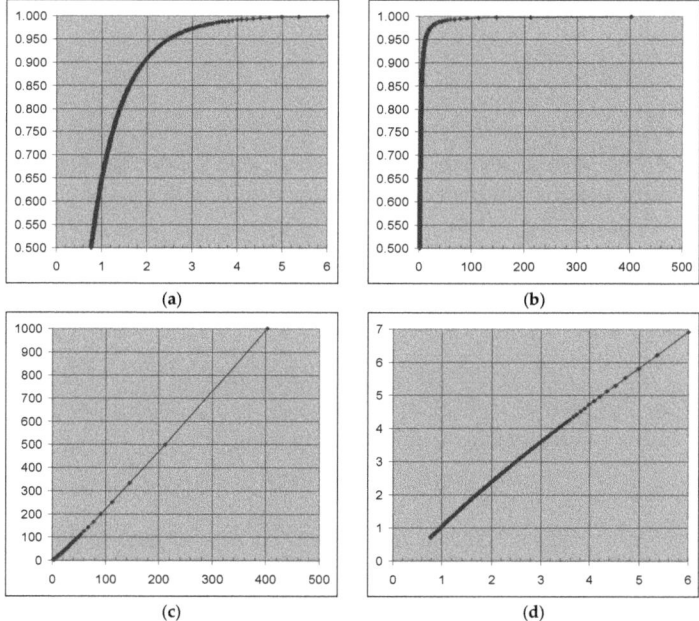

Figure 1. Probability as function of the AD statistic for a selected case (n = 25) in the Monte Carlo experiment: (**a**) $p = p(AD)$; (**b**) $p = p(e^{AD})$; (**c**) α-1 vs. eAD; (**d**) $-\ln(\alpha)$ vs. AD.

A simple linear regression $y \sim \hat{y} = a \cdot x + b$ for $x \leftarrow e^{AD}$ and $y \leftarrow \alpha - 1 = 1/(1 - p)$ will do most of the job for providing the values of α associated with the values of the AD. Since the dependence is almost linear, polynomial or rational functions will perform worse, as proven in the tests. A better alternative is to feed the model with fractional powers of x. By doing this, the bigger numbers will not be disfavored (square root of 100 is 10, which is ten times lower than 100, while square root of 1 is 1; thus, the weight of the linear component is less affected for bigger numbers). On the other hand, looking to the AD definition, the probability is raised at a variable power, and therefore, to turn back to it, in the conventional sense of operation, is to do root. Our proposed model is given in Equation (3):

$$\hat{y} = a_0 + a_1 x^{1/4} + a_2 x^{2/4} + a_3 x^{3/4} + a_4 x \tag{3}$$

The statistics associated with the proposed model for data presented in Figure 1 are given in Table 6.

Table 6. Proposed model tested for the AD = AD(p) series for $n = 25$. SST: Sum of Squares: Total; SSRes: Sum of Squares: Residuals; SSE = Sum of Squares Error.

Coefficient	Value (95% CI)	SE	t-Value
a_0	4.160 (4.126 to 4.195)	0.017567	237
a_1	−10.327 (−10.392 to −10.263)	0.032902	−314
a_2	9.357 (9.315 to 9.400)	0.02178	430
a_3	−6.147 (−6.159 to −6.135)	0.00601	−1023
a_4	3.4925 (3.4913 to 3.4936)	0.000583	5993
SST = 1550651, SSRes = 0.0057, SSE = 0.0034, r^2_{adj} = 0.999999997			

The analysis of the results presented in Table 6 showed that all coefficients are statistically significant, and their significance increases from the coefficient of $AD^{1/4}$ to the coefficient of the AD. Furthermore, the residuals of the regression are with ten orders of magnitude less than the total residuals (F value = 3.4×10^{10}). The adjusted determination coefficient has eight consecutive nines.

The model is not finished yet, because we need a model that also embeds the sample size (n). Inverse powers of n are the best alternatives as already suggested in the literature [43]. Therefore, for each coefficient (from a_0 to a_4), a function penalizing the small samples was used similarly:

$$\hat{a}_i = b_{0,i} + b_{1,i} n^{-1} + b_{2,i} n^{-2} + b_{3,i} n^{-3} + b_{4,i} n^{-4}. \tag{4}$$

With these replacements, the whole model providing the probability as a function of AD statistic and n is given by Equation (5):

$$\hat{y} = \sum_{i=0}^{4} \sum_{j=0}^{4} b_{i,j} x^{i/4} n^{-j}, \tag{5}$$

where $\hat{y} = 1/(1-p)$, $b_{i,j}$ = coefficients, $x = e^{AD}$, n = sample size.

3. Simulation Results

Twenty-five coefficients were calculated for Equation (5) from 60 values associated to sample sizes from 2 to 61, based on 500 values of p ($0.500 \leq p \leq 0.999$) and with a step of 0.001. The values of the obtained coefficients along with the related Student t-statistic are given in Table 7.

Table 7. Coefficients of the proposed model and their Student *t*-values provided in round brackets.

$b_{i,j}$ ($t_{i,j}$)	j = 0	j = 1	j = 2	j = 3	j = 4
i = 0	5.6737 (710)	−38.9087 (4871)	88.7461 (11111)	−179.5470 (22479)	199.3247 (24955)
i = 1	−13.5729 (1699)	83.6500 (10473)	−181.6768 (22746)	347.6606 (43526)	−367.4883 (46009)
i = 2	12.0750 (1512)	−70.3770 (8811)	139.8035 (17503)	−245.6051 (30749)	243.5784 (30496)
i = 3	−7.3190 (916)	30.4792 (3816)	−49.9105 (6249)	76.7476 (9609)	−70.1764 (8786)
i = 4	3.7309 (467)	−6.1885 (775)	7.3420 (919)	−9.3021 (1165)	7.7018 (964)

3.1. Stratified vs. Random

The same experiment was conducted with both simple and random stratified Mersenne Twister method [53] to assess the magnitude of the increases in the resolution of the AD statistic. The differences between the two scenarios were calculated and plotted in Figure 2.

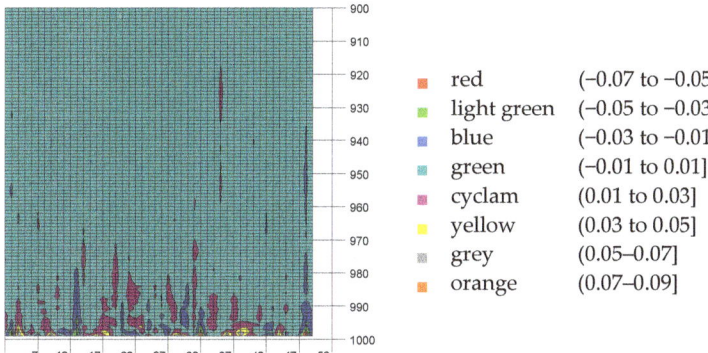

Figure 2. The effect in differences between classical and stratified random in calculated AD statistic.

3.2. Analysis of Residuals

The residuals, defined as the difference between the probability obtained by Monte Carlo simulation and the value estimated by the proposed model, without and with transformation (*ln* and respectively *log*), were analyzed. For each probability (*p* ranging from 0.500 to 0.999 with a step of 0.001; 500 values) associated with the statistic (AD) based on the MC simulation for *n* ranging from 2 to 61 (60 values), 30,000 distinct pairs (*p*, *n*, AD) were collected and investigated. The descriptive statistics of residuals are presented in Table 8.

Table 8. Residuals: descriptive statistics.

Parameter	$(p - \hat{p})$	$\ln(p - \hat{p})$	$\log(p - \hat{p})$
Arithmetic mean	3.04×10^{-7}	−18.8283	−8.17703
Standard deviation	2.55×10^{-6}	3.9477	1.7144
Standard error	1.47×10^{-8}	0.02279	0.009898
Median	1.5×10^{-8}	−18.0132	−7.82304
Mode	9.52×10^{-8}	−16.1677	−7.02156
Minimum	1.32×10^{-18}	−41.167	−17.8786
Maximum	0.000121	−9.02296	−3.9186

The most frequent value of residuals (~99%) equals with 0.000007 when no transformed data are investigated (Figure 3, left-hand graph). The right-hand chart in Figure 3 depicted the distribution of the same data, but expressed in logarithmic scale, showing a better agreement with normal distribution for the transformed residuals.

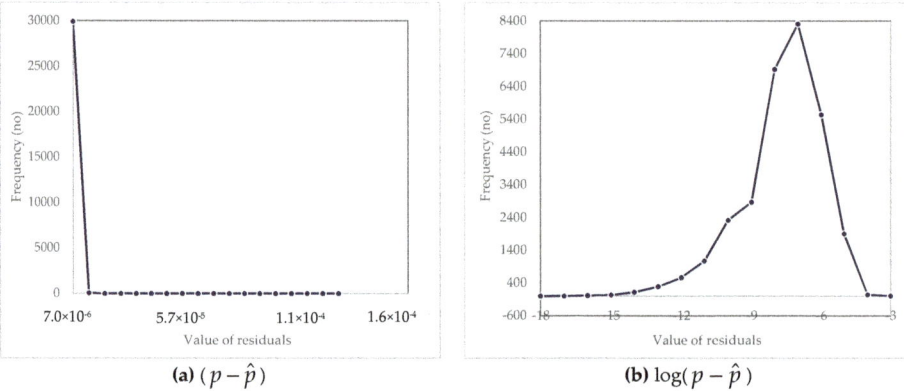

(a) $(p - \hat{p})$ (b) $\log(p - \hat{p})$

Figure 3. Distribution of residuals (differences between MC-simulated values and the values estimated by our model) for the probability from regression for the whole pool of data (30,000 pairs). (a) untransformed data (b) log transformed data

A sample of p ranging from 0.500 to 0.995 with a step of 0.005 (100 values), and for n in the same range (from 2 to 61; 60 values) was extracted from the whole pool of data, and a 3D mesh with 6000 grid points was constructed. Figure 4 represents the differences $\log_{10}(p - \hat{p})$ ($\hat{p}$ is calculated with Equation (5)) and the values of the $b_{i,j}$ coefficients given in Table 4. For convenience, the equation for $\hat{p}$ and ($\alpha \equiv 1 \times p$) are

$$\hat{p} = 1 - \left(\sum_{i=0}^{4} \sum_{j=0}^{4} b_{i,j} x^{i/4} n^{-j} \right)^{-1}$$

$$\hat{\alpha} = \left(\sum_{i=0}^{4} \sum_{j=0}^{4} b_{i,j} x^{i/4} n^{-j} \right)^{-1}.$$

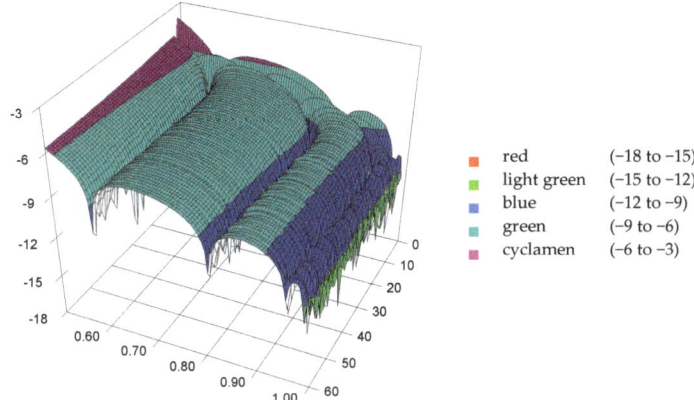

Figure 4. 3D plot of the estimation error for data expressed in logarithm scale as function of p (ranging from 0.500 to 0.999) and n (ranging from 2 to 61).

Figure 4 reveals that the calculated Equation (5) and the expected values (from MC simulation for AD = AD(p,n)) differ less than 1‰ (−3 on the top of the Z axis). Even more than that, with departure from $n = 2$, and from $p = 0.500$ to $n = 61$, or to $p = 0.999$, the difference dramatically decreases to 10^{-6} (visible on the Z-axis as −6 moving from $n = 2$ to $n = 61$), to 10^{-9} (visible on the plot visible on X-axis as −9 moving from $p = 0.500$ to $p = 0.995$), and even to 10^{-15} (visible on the plot on Z-axis as −15 moving on both from $p = 0.500$ to $p = 0.995$ and from $n = 2$ to $n = 61$). This behavior shows that the model was designed in a way in which the estimation error ($p − \hat{p}$) would be minimal for small α (α close to 0; p close to 1). A regular half-circle shape pattern, depicted in Figure 4, suggests that an even more precise method than the one archived by the proposed model must be done with periodic functions.

Figure 5 illustrates, more obviously, this pattern with the peak at $n = 2$ and $p = 0.500$.

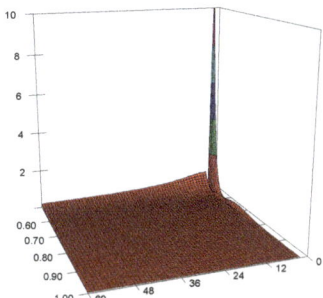

Figure 5. 3D plot of the estimation error for untransformed data: Z-axis show the $10^5 \cdot (p − \hat{p})$ as a function of p (ranging from 0.500 to 0.999) and n (ranging from 2 to 61).

Median of residuals expressed in logarithmic scale indicate that half of the points have exactly seven digits (e.g., 0.98900000 vs. 0.98900004). The cumulative frequencies for the residuals represented in logarithmic scale also show that 75% have exactly six digits, while over 99% have exactly five digits. The agreement between the observed Monte Carlo and the regression model is excellent ($r^2(n = 30{,}000) = 0.99999$) with a minimum value for the sum of squares of residuals (0.002485). These results sustain the validity of the proposed model.

4. Case Study

Twenty sets of experimental data (Table 9) were used to test the hypothesis of the normal distribution:

H_1: The distribution of experimental data is not significantly different from the theoretical normal distribution.

H_2: The distribution of experimental data is not significantly different from the theoretical normal distribution.

Table 9. Characteristics of the investigated datasets.

Set ID	What the Data Represent?	Sample Size	Reference
1	Distance (m) on treadmill test, applied on subject ts with peripheral arterial disease	24	[54]
2	Waist/hip ratio, determined in obese insulin-resistant patients	53	[55]
3	Insulin-like growth factor 2 (pg/mL) on newborns	60	[56]
4	Chitotriosidase activity (nmol/mL/h) on patients with critical limb ischemia	43	[57]
5	Chitotriosidase activity (nmol/mL/h) on patients with critical limb ischemia and on controls	86	[57]
6	Total antioxidative capacity (Eq/L) on the control group	10	[58]
7	Total antioxidative capacity (Eq/L) on the group with induced migraine	40	[53]
8	Mini mental state examination score (points) elderly patients with cognitive dysfunction	163	[59]
9	Myoglobin difference (ng/mL) (postoperative–preoperative) in patients with total hip arthroplasty	70	[60]
10	The inverse of the molar concentration of carboquinone derivatives, expressed in logarithmic scale	37	[61]
11	Partition coefficient expressed in the logarithmic scale of flavonoids	40	[62]
12	Evolution of determination coefficient in the identification of optimal model for lipophilicity of polychlorinated biphenyls using a genetic algorithm	30	[63]
13	Follow-up days in the assessment of the clinical efficiency of a vaccine	31	[64]
14	Strain ratio elastography to cervical lymph nodes	50	[65]
15	Total strain energy (eV) of C_{42} fullerene isomers	45	[66]
16	Breslow index (mm) of melanoma lesions	29	[67]
17	Determination coefficient distribution in full factorial analysis on one-cage pentagonal face C_{40} congeners: dipole moment	44	[68]
18	The concentration of spermatozoids (millions/mL) in males with ankylosing spondylitis	60	[69]
19	The parameter of the Poisson distribution	31	[70]
20	Corolla diameter of *Calendula officinalis* L. for Bon-Bon Mix × Bon-Bon Orange	28	[71]

Experimental data were analyzed with EasyFit Professional (v. 5.2) [72], and the retrieved AD statistic, along with the conclusion of the test (Reject H_0?) at a significance level of 5% were recorded.

The AD statistic and the sample size for each dataset were used to retrieve the *p*-value calculated with our method. As a control method, the formulas presented in Table 3 [43], implemented in an Excel file (SPC for Excel) [47], were used. The obtained results are presented in Table 10.

Table 10. Anderson–Darling (AD) statistic, associated *p*-values, and test conclusion: comparisons.

Set	EasyFit		Our Method		SPC for Excel	
	AD Statistic	Reject H_0?	*p*-Value	Reject H_0?	*p*-Value	Reject H_0?
1	1.18	No	0.2730	No	0.0035	Yes
2	1.34	No	0.2198	No	0.0016	Yes
3	15.83	Yes	3.81×10^{-8}	Yes	0.0000	Yes
4	1.59	No	0.1566	No	4.63×10^{-15}	Yes
5	6.71	Yes	0.0005	Yes	1.44×10^{-16}	Yes
6	0.18	No	o.o.r.		0.8857	No
7	3.71	Yes	0.0122	Yes	1.93×10^{-9}	Yes
8	11.70	Yes	2.49×10^{-6}	Yes	3.45×10^{-28}	Yes
9	0.82	No	0.4658	No	0.0322	Yes
10	0.60	No	0.6583	No	0.1109	No
11	0.81	No	0.4752	No	0.0334	Yes
12	0.34	No	o.o.r.		0.4814	No
13	4.64	Yes	0.0044	Yes	0.0000	Yes
14	1.90	No	0.1051	No	0.0001	Yes
15	0.39	No	0.9297	No	0.3732	No
16	0.67	No	0.5863	No	0.0666	No
17	5.33	Yes	0.0020	Yes	2.23×10^{-13}	Yes
18	2.25	No	0.0677	No	9.18×10^{-6}	Yes
19	1.30	No	0.2333	No	0.0019	Yes
20	0.58	No	0.6774	No	0.1170	No

AD = Anderson–Darling; o.o.r = out of range.

A perfect concordance was observed in regard to the statistical conclusion regarding the normal distribution, when our method was compared to the judgment retrieved by EasyFit. The concordance of the results between SPC and EasyFit, respectively, with the proposed method, was 60%, with discordant results for both small (e.g., *n* = 24, set 1) samples as well as high (e.g., *n* = 70, set 9) sample sizes. Normal probability plots (P–P) and the quantile–quantile plots (Q–Q) of these sets show slight, but not significant deviations from the expected normal distribution (Figure 6).

Without any exceptions, the *p*-values calculated by our implemented method had higher values compared to the *p*-values achieved by SPC for Excel. The most substantial difference is observed for the largest dataset (set 8), while the smallest difference is noted for the set with 45 experimental data values (set 15). The lowest *p*-value was obtained by the implemented method for set 3 (see Table 10); the SPC for Excel retrieves, for this dataset, a value of 0.0000. The next smallest *p*-value was observed for set 8. For both these sets, an agreement related to the statistical decision was found (see Table 10).

Our team has previously investigated the effect of sample size on the probability of Anderson–Darling test, and the results are published online at http://l.academicdirect.org/Statistics/tests/AD/. The method proposed in this manuscript, as compared to the previous one, assures a higher resolution expressed by the lower unexplained variance between the AD and the model using a formula with a smaller number of coefficients. Furthermore, the unexplained variance of the method present in this manuscript has much less weight for big "*p*-values", and much higher weight for small "*p*-values", which means that is more appropriate to be used for low (e.g., $p \sim 10^{-5}$) and very low ($p \sim 10^{-10}$) probabilities.

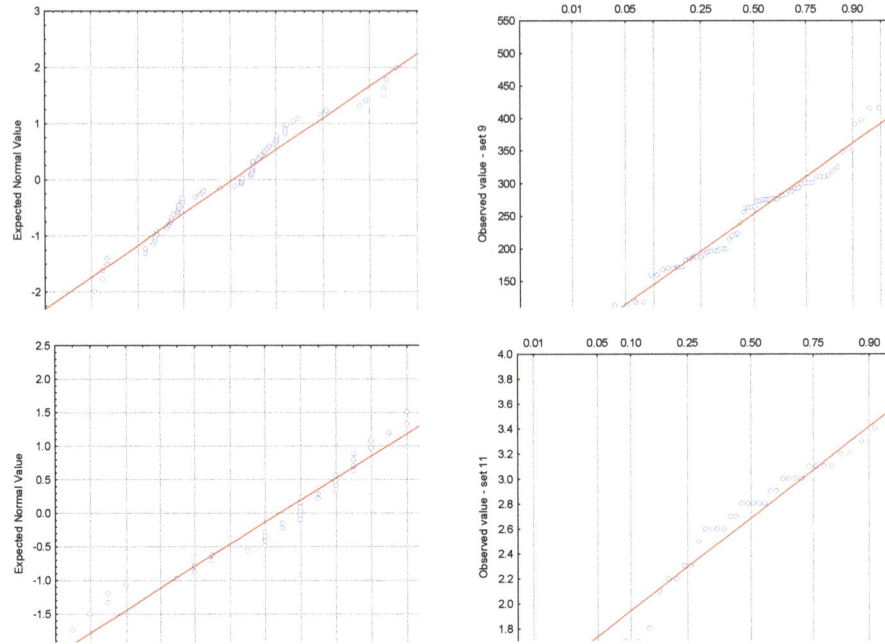

Figure 6. Normal probability plots (P–P) and quantile-quantile plot (Q–Q) by example: graphs for set 9 ($n = 70$) in the first row, and for set 11 ($n = 40$) in the second row.

Further research could be done in both the extension of the proposed method and the evaluation of its performances. The performances of the reported method could be evaluated for the whole range of sample sizes if proper computational resources exist. Furthermore, the performance of the implementation could be assessed using game theory and game experiments [73,74] using or not using diagnostic metrics (such as validation, confusion matrices, ROC analysis, analysis of errors, etc.) [75,76].

The implemented method provides a solution to the calculation of the p-values associated with Anderson–Darling statistics, giving proper weight to the sample size of the investigated experimental data. The advantage of the proposed estimation method, Equation (5), is its very low residual (unexplained variance) and its very high estimation accuracy at convergence (with increasing of in and for p near 1). The main disadvantage is related to its out of range p-values for small AD values, but an extensive simulation study could solve this issue. The worst performances of the implemented methods are observed when simultaneously n is very low (2 or 3) and p is near 0.5 (50–50%).

Author Contributions: L.J. and S.D.B. conceived and designed the experiments; L.J. performed the experiments; L.J. and S.D.B. analyzed the data; S.D.B. wrote the paper and L.J. critically reviewed the manuscript.

Acknowledgments: No grants have been received in support of the research work reported in this manuscript. No funds were received for covering the costs to publish in open access.

Conflicts of Interest: The authors declare no conflict of interest.

References

1. Nimon, K.F. Statistical assumptions of substantive analyses across the General Linear model: A Mini-Review. *Front. Psychol.* **2012**, *3*, 322. [CrossRef] [PubMed]
2. Hoekstra, R.; Kiers, H.A.; Johnson, A. Are assumptions of well-known statistical techniques checked, and why (not)? *Front. Psychol.* **2012**, *3*, 137. [CrossRef] [PubMed]
3. Kolmogorov, A. Sulla determinazione empirica di una legge di distribuzione. *Giornale dell'Istituto Italiano degli Attuari* **1933**, *4*, 83–91.
4. Smirnov, N. Table for estimating the goodness of fit of empirical distributions. *Ann. Math. Stat.* **1948**, *19*, 279–281. [CrossRef]
5. Anderson, T.W.; Darling, D.A. Asymptotic theory of certain "goodness-of-fit" criteria based on stochastic processes. *Ann. Math. Stat.* **1952**, *23*, 193–212. [CrossRef]
6. Anderson, T.W.; Darling, D.A. A Test of Goodness-of-Fit. *J. Am. Stat. Assoc.* **1954**, *49*, 765–769. [CrossRef]
7. Pearson, K. Contribution to the mathematical theory of evolution. II. Skew variation in homogenous material. *Philos. Trans. R. Soc. Lond.* **1895**, *91*, 343–414. [CrossRef]
8. Cramér, H. On the composition of elementary errors. *Scand. Actuar. J.* **1928**, *1*, 13–74. [CrossRef]
9. Von Mises, R.E. *Wahrscheinlichkeit, Statistik und Wahrheit*; Julius Springer: Berlin, Germany, 1928.
10. Shapiro, S.S.; Wilk, M.B. An analysis of variance test for normality (complete samples). *Biometrika* **1965**, *52*, 591–611. [CrossRef]
11. Jarque, C.M.; Bera, A.K. Efficient tests for normality, homoscedasticity and serial independence of regression residuals. *Econ. Lett.* **1980**, *6*, 255–259. [CrossRef]
12. Jarque, C.M.; Bera, A.K. Efficient tests for normality, homoscedasticity and serial independence of regression residuals: Monte Carlo evidence. *Econ. Lett.* **1981**, *7*, 313–318. [CrossRef]
13. Jarque, C.M.; Bera, A.K. A test for normality of observations and regression residuals. *Int. Stat. Rev.* **1987**, *55*, 163–172. [CrossRef]
14. D'Agostino, R.B.; Belanger, A.; D'Agostino, R.B., Jr. A suggestion for using powerful and informative tests of normality. *Am. Stat.* **1990**, *44*, 316–321. [CrossRef]
15. Lilliefors, H.W. On the Kolmogorov-Smirnov test for normality with mean and variance unknown. *J. Am. Stat. Assoc.* **1967**, *62*, 399–402. [CrossRef]
16. Van Soest, J. Some experimental results concerning tests of normality. *Stat. Neerl.* **1967**, *21*, 91–97. [CrossRef]
17. Jänstchi, L.; Bolboacă, S.D. Performances of Shannon's entropy statistic in assessment of distribution of data. *Ovidius Univ. Ann. Chem.* **2017**, *28*, 30–42. [CrossRef]
18. Noughabi, H.A. Two Powerful Tests for Normality. *Ann. Data Sci.* **2016**, *3*, 225–234. [CrossRef]
19. Zamanzade, E.; Arghami, N.R. Testing normality based on new entropy estimators. *J. Stat. Comput. Simul.* **2012**, *82*, 1701–1713. [CrossRef]
20. Peng, H.; Tan, F. Jackknife empirical likelihood goodness-of-fit tests for U-statistics based general estimating equations. *Bernoulli* **2018**, *24*, 449–464. [CrossRef]
21. Shah, R.D.; Bühlmann, P. Goodness-of-fit tests for high dimensional linear models. Journal of the Royal Statistical Society. *Ser. B Stat. Methodol.* **2018**, *80*, 113–135. [CrossRef]
22. Balakrishnan, K.; Sooriyarachchi, M.R. A goodness of fit test for multilevel survival data. *Commun. Stat. Simul. Comput.* **2018**, *47*, 30–47. [CrossRef]
23. Perera, A.A.P.N.M.; Sooriyarachchi, M.R.; Wickramasuriya, S.L. A Goodness of Fit Test for the Multilevel Logistic Model. *Commun. Stat. Simul. Comput.* **2016**, *45*, 643–659. [CrossRef]
24. Villaseñor, J.A.; González-Estrada, E.; Ochoa, A. On Testing the inverse Gaussian distribution hypothesis. *Sankhya B* 2017. [CrossRef]
25. MacKenzie, D.W. Applying the Anderson-Darling test to suicide clusters: Evidence of contagion at U. S. Universities? *Crisis* **2013**, *34*, 434–437. [CrossRef] [PubMed]
26. Müller, C.; Kloft, H. Parameter estimation with the Anderson-Darling test on experiments on glass. *Stahlbau* **2015**, *84*, 229–240. [CrossRef]
27. İçen, D.; Bacanlı, S. Hypothesis testing for the mean of inverse Gaussian distribution using α-cuts. *Soft Comput.* **2015**, *19*, 113–119. [CrossRef]
28. Ghasemi, A.; Zahediasl, S. Normality tests for statistical analysis: A guide for non-statisticians. *Int. J. Endocrinol. Metab.* **2012**, *10*, 486–489. [CrossRef] [PubMed]

29. Hwe, E.K.; Mohd Yusoh, Z.I. Validation guideline for small scale dataset classification result in medical domain. *Adv. Intell. Syst. Comput.* **2018**, *734*, 272–281. [CrossRef]
30. Ruxton, G.D.; Wilkinson, D.M.; Neuhäuser, M. Advice on testing the null hypothesis that a sample is drawn from a normal distribution. *Anim. Behav.* **2015**, *107*, 249–252. [CrossRef]
31. Lang, T.A.; Altman, D.G. Basic statistical reporting for articles published in biomedical journals: The "Statistical Analyses and Methods in the Published Literature" or The SAMPL Guidelines. In *Science Editors' Handbook*; European Association of Science Editors, Smart, P., Maisonneuve, H., Polderman, A., Eds.; EASE: Paris, France, 2013; Available online: http://www.equator-network.org/wp-content/uploads/2013/07/SAMPL-Guidelines-6-27-13.pdf (accessed on 3 January 2018).
32. Curran-Everett, D.; Benos, D.J. American Physiological Society. Guidelines for reporting statistics in journals published by the American Physiological Society.
33. Curran-Everett, D.; Benos, D.J. Guidelines for reporting statistics in journals published by the American Physiological Society: The sequel. *Adv. Physiol. Educ.* **2007**, *31*, 295–298. [CrossRef] [PubMed]
34. Razali, N.M.; Wah, Y.B. Power comparison of Shapiro-Wilk, Kolmogorov-Smirnov, Lilliefors and Anderson-Darling tests. *J. Stat. Model. Anal.* **2011**, *2*, 21–33.
35. Tui, I. Normality Testing—A New Direction. *Int. J. Bus. Soc. Sci.* **2011**, *2*, 115–118.
36. Saculinggan, M.; Balase, E.A. Empirical Power Comparison of Goodness of Fit Tests for Normality in the Presence of Outliers. *J. Phys. Conf. Ser.* **2013**, *435*, 012041. [CrossRef]
37. Sánchez-Espigares, J.A.; Grima, P.; Marco-Almagro, L. Visualizing type II error in normality tests. *Am. Stat.* **2017**. [CrossRef]
38. Yap, B.W.; Sim, S.H. Comparisons of various types of normality tests. *J. Stat. Comput. Simul.* **2011**, *81*, 2141–2155. [CrossRef]
39. Patrício, M.; Ferreira, F.; Oliveiros, B.; Caramelo, F. Comparing the performance of normality tests with ROC analysis and confidence intervals. *Commun. Stat. Simul. Comput.* **2017**, *46*, 7535–7551. [CrossRef]
40. Mbah, A.K.; Paothong, A. Shapiro-Francia test compared to other normality test using expected p-value. *J. Stat. Comput. Simul.* **2015**, *85*, 3002–3016. [CrossRef]
41. Arshad, M.; Rasool, M.T.; Ahmad, M.I. Anderson Darling and Modified Anderson Darling Tests for Generalized Pareto Distribution. *Pak. J. Appl. Sci.* **2003**, *3*, 85–88.
42. Stephens, M.A. Goodness of fit for the extreme value distribution. *Biometrika* **1977**, *64*, 585–588. [CrossRef]
43. D'Agostino, R.B.; Stephens, M.A. *Goodness-of-Fit Techniques*; Marcel-Dekker: New York, NY, USA, 1986; pp. 123, 146.
44. Shin, H.; Jung, Y.; Jeong, C.; Heo, J.-H. Assessment of modified Anderson–Darling test statistics for the generalized extreme value and generalized logistic distributions. *Stoch. Environ. Res. Risk Assess.* **2012**, *26*, 105–114. [CrossRef]
45. De Micheaux, P.L.; Tran, V.A. PoweR: A Reproducible Research Tool to Ease Monte Carlo Power Simulation Studies for Goodness-of-fit Tests in R. *J. Stat. Softw.* **2016**, *69*. Available online: https://www.jstatsoft.org/article/view/v069i03 (accessed on 10 April 2018).
46. 6ixSigma.org—Anderson Darling Test. Available online: http://6ixsigma.org/SharedFiles/Download.aspx?pageid=14&mid=35&fileid=147 (accessed on 2 June 2017).
47. Spcforexcel. Anderson-Darling Test for Normality. 2011. Available online: http://www.spcforexcel.com/knowledge/basic-statistics/anderson-darling-test-for-normality (accessed on 2 June 2017).
48. Qimacros—Data Normality Tests Using p and Critical Values in QI Macros. © 2015 KnowWare International Inc. Available online: http://www.qimacros.com/hypothesis-testing//data-normality-test/#anderson (accessed on 2 June 2017).
49. Jäntschi, L.; Bolboacă, S.D. Distribution Fitting 2. Pearson-Fisher, Kolmogorov-Smirnov, Anderson-Darling, Wilks-Shapiro, Kramer-von-Misses and Jarque-Bera statistics. *Bull. Univ. Agric. Sci. Vet. Med. Cluj-Napoca Hortic.* **2009**, *66*, 691–697.
50. Mosteller, F. Questions and Answers—Combining independent tests of significance. *Am. Stat.* **1948**, *2*, 30–31. [CrossRef]
51. Bolboacă, S.D.; Jäntschi, L.; Sestraş, A.F.; Sestraş, R.E.; Pamfil, D.C. Pearson-Fisher Chi-Square Statistic Revisited. *Information* **2011**, *2*, 528–545. [CrossRef]
52. Rahman, M.; Pearson, L.M.; Heien, H.C. A Modified Anderson-Darling Test for Uniformity. *Bull. Malays. Math. Sci. Soc.* **2006**, *29*, 11–16.

53. Matsumoto, M.; Nishimura, T. Mersenne twister: A 623-dimensionally equidistributed uniform pseudo-random number generator (PDF). *ACM Trans. Model. Comput. Simul.* 1998, *8*, 3–30. [CrossRef]
54. Ciocan, A.; Ciocan, R.A.; Gherman, C.D.; Bolboacă, S.D. Evaluation of Patients with Lower Extremity Peripheral Artery Disease by Walking Tests: A Pilot Study. *Not. Sci. Biol.* 2017, *9*, 473–479. [CrossRef]
55. Răcătăianu, N.; Bolboacă, S.D.; Sitar-Tăut, A.-V.; Marza, S.; Moga, D.; Valea, A.; Ghervan, C. The effect of Metformin treatment in obese insulin-resistant patients with euthyroid goiter. *Acta Clin. Belg. Int. J. Clin. Lab. Med.* 2018. [CrossRef] [PubMed]
56. Hășmășanu, M.G.; Baizat, M.; Procopciuc, L.M.; Blaga, L.; Văleanu, M.A.; Drugan, T.C.; Zaharie, G.C.; Bolboacă, S.D. Serum levels and ApaI polymorphism of insulin-like growth factor 2 on intrauterine growth restriction infants. *J. Matern.-Fetal Neonatal Med.* 2018, *31*, 1470–1476. [CrossRef] [PubMed]
57. Ciocan, R.A.; Drugan, C.; Gherman, C.D.; Cătană, C.-S.; Ciocan, A.; Drugan, T.C.; Bolboacă, S.D. Evaluation of Chitotriosidase as a Marker of Inflammatory Status in Critical Limb Ischemia. *Ann. Clin. Lab. Sci.* 2017, *47*, 713–719. [PubMed]
58. Bulboacă, A.E.; Bolboacă, S.D.; Stănescu, I.C.; Sfrângeu, C.-A.; Bulboacă, A.C. Preemptive Analgesic and Anti-Oxidative Effect of Curcumin for Experimental Migraine. *BioMed Res. Int.* 2017, *2017*, 4754701. [CrossRef]
59. Bulboacă, A.E.; Bolboacă, S.D.; Bulboacă, A.C.; Prodan, C.I. Association between low thyroid-stimulating hormone, posterior cortical atrophy and nitro-oxidative stress in elderly patients with cognitive dysfunction. *Arch. Med. Sci.* 2017, *13*, 1160–1167. [CrossRef] [PubMed]
60. Nistor, D.-V.; Caterev, S.; Bolboacă, S.D.; Cosma, D.; Lucaciu, D.O.G.; Todor, A. Transitioning to the direct anterior approach in total hip arthroplasty. Is it a true muscle sparing approach when performed by a low volume hip replacement surgeon? *Int. Orthopt.* 2017, *41*, 2245–2252. [CrossRef] [PubMed]
61. Bolboacă, S.D.; Jäntschi, L. Comparison of QSAR Performances on Carboquinone Derivatives. *Sci. World J.* 2009, *9*, 1148–1166. [CrossRef] [PubMed]
62. Harsa, A.M.; Harsa, T.E.; Bolboacă, S.D.; Diudea, M.V. QSAR in Flavonoids by Similarity Cluster Prediction. *Curr. Comput.-Aided Drug Des.* 2014, *10*, 115–128. [CrossRef] [PubMed]
63. Jäntschi, L.; Bolboacă, S.D.; Sestraș, R.E. A Study of Genetic Algorithm Evolution on the Lipophilicity of Polychlorinated Biphenyls. *Chem. Biodivers.* 2010, *7*, 1978–1989. [CrossRef] [PubMed]
64. Chirilă, M.; Bolboacă, S.D. Clinical efficiency of quadrivalent HPV (types 6/11/16/18) vaccine in patients with recurrent respiratory papillomatosis. *Eur. Arch. Oto-Rhino-Laryngol.* 2014, *271*, 1135–1142. [CrossRef] [PubMed]
65. Lenghel, L.M.; Botar-Jid, C.; Bolboacă, S.D.; Ciortea, C.; Vasilescu, D.; Băciuț, G.; Dudea, S.M. Comparative study of three sonoelastographic scores for differentiation between benign and malignant cervical lymph nodes. *Eur. J. Radiol.* 2015, *84*, 1075–1082. [CrossRef] [PubMed]
66. Bolboacă, S.D.; Jäntschi, L. Nano-quantitative structure-property relationship modeling on C42 fullerene isomers. *J. Chem.* 2016, *2016*, 1791756. [CrossRef]
67. Botar-Jid, C.; Cosgarea, R.; Bolboacă, S.D.; Șenilă, S.; Lenghel, M.L.; Rogojan, L.; Dudea, S.M. Assessment of Cutaneous Melanoma by Use of Very- High-Frequency Ultrasound and Real-Time Elastography. *Am. J. Roentgenol.* 2016, *206*, 699–704. [CrossRef] [PubMed]
68. Jäntschi, L.; Balint, D.; Pruteanu, L.L.; Bolboacă, S.D. Elemental factorial study on one-cage pentagonal face nanostructure congeners. *Mater. Discov.* 2016, *5*, 14–21. [CrossRef]
69. Micu, M.C.; Micu, R.; Surd, S.; Girlovanu, M.; Bolboacă, S.D.; Ostensen, M. TNF-a inhibitors do not impair sperm quality in males with ankylosing spondylitis after short-term or long-term treatment. *Rheumatology* 2014, *53*, 1250–1255. [CrossRef] [PubMed]
70. Sestraș, R.E.; Jäntschi, L.; Bolboacă, S.D. Poisson Parameters of Antimicrobial Activity: A Quantitative Structure-Activity Approach. *Int. J. Mol. Sci.* 2012, *13*, 5207–5229. [CrossRef] [PubMed]
71. Bolboacă, S.D.; Jäntschi, L.; Baciu, A.D.; Sestraș, R.E. Griffing's Experimental Method II: Step-By-Step Descriptive and Inferential Analysis of Variances. *JP J. Biostat.* 2011, *6*, 31–52.
72. EasyFit. MathWave Technologies. Available online: http://www.mathwave.com (accessed on 25 March 2018).
73. Arena, P.; Fazzino, S.; Fortuna, L.; Maniscalco, P. Game theory and non-linear dynamics: The Parrondo Paradox case study. *Chaos Solitons Fractals* 2003, *17*, 545–555. [CrossRef]

74. Ergün, S.; Aydoğan, T.; Alparslan Gök, S.Z. A Study on Performance Evaluation of Some Routing Algorithms Modeled by Game Theory Approach. *AKU J. Sci. Eng.* **2016**, *16*, 170–176.
75. Hossin, M.; Sulaiman, M.N. A review on evaluation metrics for data classification evaluations. *Int. J. Data Min. Knowl. Manag. Process* **2015**, *5*, 1–11. [CrossRef]
76. Gopalakrishna, A.K.; Ozcelebi, T.; Liotta, A.; Lukkien, J.J. Relevance as a Metric for Evaluating Machine Learning Algorithms. In *Machine Learning and Data Mining in Pattern Recognition*; Perner, P., Ed.; Lecture Notes in Computer Science; Springer: Berlin/Heidelberg, Germany, 2013; Volume 7988.

© 2018 by the authors. Licensee MDPI, Basel, Switzerland. This article is an open access article distributed under the terms and conditions of the Creative Commons Attribution (CC BY) license (http://creativecommons.org/licenses/by/4.0/).

Article

Optimal Repeated Measurements for Two Treatment Designs with Dependent Observations: The Case of Compound Symmetry

Miltiadis S. Chalikias

Department of Accounting and Finance, School of Business, Economics and Social Sciences, University of West Attica, 12244 Egaleo, Greece; mchalikias@hotmail.com or mchalik@uniwa.gr

Received: 18 February 2019; Accepted: 13 April 2019; Published: 25 April 2019

Abstract: In this paper, we construct optimal repeated measurement designs of two treatments for estimating direct effects, and we examine the case of compound symmetry dependency. We present the model and the design that minimizes the variance of the estimated difference of the two treatments. The optimal designs with dependent observations in a compound symmetry model are the same as in the case of independent observations.

Keywords: repeated measurement designs; compound symmetry

1. Introduction

In repeated measurement designs, a sequence of treatments is applied to each experimental unit (e.u.). In particular, one treatment is applied in each period. For example, for two treatments, A and B, and three periods, a possible sequence is ABA, which means that the treatments A, B, and A are respectively applied at the beginning of each of the three periods. The direct effect of a treatment is the effect of the treatment which is applied in the period that is examined. The residual effect is the effect of the treatment which is applied in the period preceding the period that is examined. In the case of two treatments, A and B, the direct τ_A and τ_B can be estimated. In every period, a treatment is applied, so either τ_A or τ_B is estimable. In this paper, the parameter of interest is the difference of direct effects $\tau = \tau_A - \tau_B$.

Most researchers who have investigated repeated measurement designs, such as [1–6], have been occupied with universally optimal designs where the observations are independent. However, researchers have also shown interest in designs with dependent observations, as in the cases of [7–11].

The model we use in this paper, and which is presented below, was first introduced by Hedayat and Afsarinejad [12,13]. In previous research [14,15] using this model, the author of this article studied two treatment designs under the assumption that consecutive observations were independent. Building on that previous work, in the present article the author examines the case of compound symmetry dependency. The aim is to find a design that corresponds to a minimum variance estimator.

2. The Model

A compound symmetry model has the following characteristics:

(i) For each sequence, the variance matrix is of the form $\Sigma_m = a\mathbf{I}_m + b\mathbf{J}_m$, where $\mathbf{I}_m$ is the unit $m \times m$ matrix, and $\mathbf{J}_m$ is the $m \times m$ matrix where all elements are equal to 1 (m is the number of periods).

(ii) The observations corresponding to different treatment sequences (different e.u.) are independent, and the number of sequences is 2^m.

The goal is to find the design that corresponds to the minimum variance estimator. I show that, in this case, the optimal design regarding the direct effect is the same as in the model of independent observations, and only the variance of the estimator is different.

The model is [12]:

$$y_{ijk} = \mu + \tau + \pi_j + \delta_{i,j-1} + \gamma_i + \zeta_k + e_{ijk} \tag{1}$$

j corresponds to the j-th period, $j = 1, 2, \ldots, m$;
i corresponds to the i-th sequence, $i = 0, 1, \ldots 2^m - 1$;
k corresponds to the unit $k = 1, 2, \ldots, n$;
τ_A, τ_B: are direct effects of treatments A and B;
π_j: is the effect of the j-th period;
δ_A, δ_B: are the residual effects of A and B;
γ_i: is the effect of the i-th sequence; and
ζ_k: is the effect of the k-th e.u. (subject effect), which is a random variable, independent of the error e_{ijk}.

The errors e_{ijk} are assumed to be independent. However, the quantities $\zeta_k + e_{ijk}$ are independent only between sequences and not within sequences.

The overparameterized model vector form of the above model is written as:

$$Y = \tau_A \tau_A + \tau_B \tau_B + \delta_A \delta_A + \delta_B \delta_B + \pi_1 \pi_1 + \cdots + \pi_m \pi_m + + \gamma_0 \gamma_0 + \cdots + \gamma_q \gamma_q + e \tag{2}$$

where $q = 2^m - 1$ and $Y, \tau_A, \tau_B, \delta_A, \delta_B, \pi_1, \cdots \pi_m, \gamma_0, \cdots \gamma_q, e$ are $1 \times mn$ vectors; the direct effect vector is 1 if the treatment is A, and zero if it is B. For example, for the sequence ABB $\ldots$, $\tau_A = \begin{bmatrix} 1 \\ 0 \\ 0 \\ \vdots \end{bmatrix}$

and, in the same way, $\tau_B, \delta_A, \delta_B \pi_1$ and γ_1 are defined so that $\tau_A + \tau_B = 1_{mn}$, $\delta_A + \delta_B + \pi_1 = 1_{mn}$, and $\pi_1 + \pi_2 + \ldots + \pi_m = 1_{mn}$. Also, 1 when the ith unit is employed, and 0 elsewhere, so $\gamma_0 + \gamma_1 + \gamma_2 + \ldots + \gamma_{2^m-1} = 1_{mn}$. So, in equation (2) there are linearly dependent vectors.

Keeping only the linear independent vector [16], the model (2) is transformed to

$$E(Y) = \tau(\tau_A - \tau_B) + \delta(\delta_A - \delta_B) + \pi_1 \pi_1 + \cdots + \pi_{m-1} \pi_{m-1} + + \gamma_0 \gamma_0 + \cdots + \gamma_{q-1} \gamma_{q-1}$$

where $q = 2^m - 1$. In a vector form:

$$Y = Xb + e \Leftrightarrow Y = \begin{pmatrix} X_1 & X_2 \end{pmatrix} \begin{pmatrix} b_1 \\ b_2 \end{pmatrix} \tag{3}$$

where Y is $(mn) \times 1$, the design matrix X is $(mn) \times s$, b is $s \times 1$, e is $(mn) \times 1$, and s is the number of unknown parameters. If we are interested only in some and not in all of the parameters, then we write $b' = \begin{pmatrix} b'_1 & b'_2 \end{pmatrix}$, where b_1 is the r parameters of interest, and b_2 is the s-r remaining parameters.

We assume only one parameter of interest for the difference of the direct effects, $\tau = \tau_A - \tau_B$, which can be considered as the direct effect of A in the case of $\tau_B = 0$. In order to guarantee the estimability of the model, we postulate the restrictions $\tau_B = 0, \pi_m = 0, \gamma_{2^m-1} = 0$.

The matrix X_1 corresponds to the coefficients of τ, and the matrix X_2 corresponds to the coefficients of the rest of the non-random variables. Let us assume $V = X_2(X_2^T \Sigma^{-1} X_2)^{-1} X_2^T$ is a $(mn) \times (mn)$ matrix, P the projection matrix of X_2, $P = X_2(X_2^T X_2)^{-1} X_2^T$ and Σ are the $(mn) \times (mn)$ variance matrix of the observations.

From the ordinal least-squares equations, we derive the following relation for the estimation of the main effect τ:

$$(X_1^T \Sigma^{-1} X_1 - X_1^T \Sigma^{-1} V \Sigma^{-1} X_1) \hat{\tau} = X_1^T \Sigma^{-1} (I - P \Sigma^{-1}) Y$$

We also have
$$\text{var}(\tau) = \sigma^2(\mathbf{X}_1^T\mathbf{\Sigma}^{-1}\mathbf{X}_1 - \mathbf{X}_1^T\mathbf{\Sigma}^{-1}\mathbf{V}\mathbf{\Sigma}^{-1}\mathbf{X}_1)^{-1} = \sigma^2\mathbf{Q}^{-1} \qquad (4)$$

3. The Case of Compound Symmetry

The observations are dependent within sequences with variance matrix $\mathbf{\Sigma}_m$. The observations from different sequences are independent, therefore:

$$\mathbf{\Sigma} = \begin{bmatrix} \mathbf{\Sigma}_m & 0 & \cdots & 0 \\ 0 & \mathbf{\Sigma}_m & \cdots & 0 \\ \vdots & & \ddots & \vdots \\ 0 & 0 & \cdots & \mathbf{\Sigma}_m \end{bmatrix} \text{ and } \mathbf{V} = \begin{bmatrix} \mathbf{V}_{m0} & 0 & \cdots & 0 \\ 0 & \mathbf{V}_{m1} & \cdots & 0 \\ \vdots & & \ddots & \vdots \\ 0 & 0 & \cdots & \mathbf{V}_{mq} \end{bmatrix}$$

where $q = 2^m - 1$ and $\mathbf{V}_{mj} = \mathbf{X}_{2j}(\mathbf{X}_{2j}^T\mathbf{\Sigma}_m^{-1}\mathbf{X}_{2j})^{-1}\mathbf{X}_{2j}^T$.

In order to obtain a sequence enumeration, the binary enumeration system was used, with 0 corresponding to A, and 1 to B. Thus, we obtained the enumerations $0, 1, \ldots, 2^m - 1$. For example, if we have five periods and the sequence BABBA, then this is the 13th sequence, since $BABBA \leftrightarrow 1 \cdot 2^0 + 0 \cdot 2^0 + 1 \cdot 2^2 + 1 \cdot 2^3 + 0 \cdot 2^4 = 13$. For two periods, we have four sequences, that is, $AA \leftrightarrow 0, BA \leftrightarrow 1, AB \leftrightarrow 2, BB \leftrightarrow 3$. For three periods (two treatments) we have eight sequences:

A	B	A	B	A	B	A	B
A	A	B	B	A	A	B	B
A	A	A	A	B	B	B	B
u_0	u_1	u_2	u_3	u_4	u_5	u_6	u_7

where u_i $i = 0, 2, 3, 4, 5, 6, 7$ is the number of units that received the i-th sequence of treatments. The sequences that we obtain by substituting A for B and vice versa are called dual or reversal designs. Observe that for these sequences, we obtain the enumeration $7 - i$, $i = 0, 1, 2, 3$.

Proposition 1. *For a repeated measurement design with m periods, n experimental units, and a variance matrix $\mathbf{\Sigma}$ that consists of n diagonal block matrices of the form $\mathbf{\Sigma}_m = a\mathbf{I}_m + b\mathbf{J}_m$,*

$$(\mathbf{X}_1^T\mathbf{\Sigma}^{-1}\mathbf{X}_1 - \mathbf{X}_1^T\mathbf{\Sigma}^{-1}\mathbf{V}\mathbf{\Sigma}^{-1}\mathbf{X}_1) = \frac{1}{a}(\mathbf{X}_1^T\mathbf{X}_1 - \mathbf{X}_1^T\mathbf{P}\mathbf{X}_1)$$

where $\mathbf{P} = \mathbf{X}_2(\mathbf{X}_2^T\mathbf{X}_2)^{-1}\mathbf{X}_2^T$.

Proof. Let $\widetilde{\mathbf{X}}_1 = \mathbf{\Sigma}^{-1/2}\mathbf{X}_1, \widetilde{\mathbf{X}}_2 = \mathbf{\Sigma}^{-1/2}\mathbf{X}_2$, and $\widetilde{\mathbf{Y}} = \mathbf{\Sigma}^{-1/2}\mathbf{Y}$. Then

$$(\mathbf{X}_1^T\mathbf{\Sigma}^{-1}\mathbf{X}_1 - \mathbf{X}_1^T\mathbf{\Sigma}^{-1}\mathbf{V}\mathbf{\Sigma}^{-1}\mathbf{X}_1) = \widetilde{\mathbf{X}}_1^T\widetilde{\mathbf{X}}_1 - \widetilde{\mathbf{X}}_1^T\widetilde{\mathbf{P}}\widetilde{\mathbf{X}}_1$$

where $\widetilde{\mathbf{P}} = \widetilde{\mathbf{X}}_2(\widetilde{\mathbf{X}}_2^T\widetilde{\mathbf{X}}_2)^{-1}\widetilde{\mathbf{X}}_2^T$. In other words, $\widetilde{\mathbf{P}}$ is the matrix of the orthogonal projection to $R(\widetilde{\mathbf{X}}_2)$. □

$\mathbf{X}_{1j}$ $j = 0, 1, 2 \ldots m$ is the $m \times 1$ matrix of τ in the j-th sequence, and $\mathbf{X}_{2j}$ $j = 0, 1, 2 \ldots m$ is the $mx(m + 2^m)$ matrix of the parameters $\mu, \pi_1, \pi_2, \ldots \pi_{m-1}, \delta_A, \delta_B, \gamma_1, \gamma_2, \ldots, \gamma_q$, where $q = 2^m - 1$.

For example, for three periods ($m = 3$), we have the matrices:

$$X_1 = \begin{bmatrix} X_{10}\}u_0 \\ X_{11}\}u_1 \\ \vdots \\ X_{17}\}u_7 \end{bmatrix} \text{ and } X_2 = \begin{bmatrix} X_{20} \\ X_{21} \\ \vdots \\ X_{27} \end{bmatrix}$$

For the linear space $R(\widetilde{X}_{2j})$ and for any sequence (m observations) $R(\widetilde{X}_{2j}) = R(X_{2j})$, we observe the following:

(i) The matrix $(aI_m + bJ_m)$ is positive definite, so the matrix $(aI_m + bJ_m)^{-\frac{1}{2}}$ is also positive definite, and we conclude that:

$$(aI_m + bJ_m)^{-\frac{1}{2}} = \frac{1}{a}(I_m - \frac{b}{a+bm}J_m) \Leftrightarrow (aI_m + bJ_m)^{-\frac{1}{2}} = \frac{1}{\sqrt{a}}(I_m - \delta J_m)$$

where $\delta = \frac{\sqrt{\frac{a}{a+bm}}}{m}$, $1 - \delta \cdot m > 0$, and we have $\widetilde{X}_{2j} = \frac{1}{\sqrt{a}}(I_m - \delta J_m)^{-1}X_{2j}$.

(ii) The coefficients of the general mean are 1, so $1_m \in R(X_{2j})$ and.

$$\frac{1}{\sqrt{a}}(I_m - \delta J_m) \cdot 1_m = \frac{1}{\sqrt{a+bm}}1_m \Rightarrow 1_m \in R(\widetilde{X}_{2j})$$

(iii) If z is another column vector, and $z \in R(X_{2j})$, then

$$\frac{1}{\sqrt{a}}(I_m - \delta J_m)z = \frac{1}{\sqrt{a}}(z - \delta(1_m^T z)1_m) \Rightarrow z \in R(\widetilde{X}_{2j}) \Leftrightarrow R(\widetilde{X}_{2j}) = R(X_{2j})$$

(iv) If $\widetilde{P}_m$ is the matrix of the orthogonal projection to the linear space $R(\widetilde{X}_{2j})$, then $\widetilde{P}_{mj} = P_{mj}$, where $P_{mj} = X_{2j}(X_{2j}^T X_{2j})^{-1}X_{2j}^T$ is the matrix of the orthogonal projection to $R(X_{2j})$ and $P_{mj} \cdot 1_m = 1_m \Rightarrow P_{mj} \cdot J_m = J_m$. From the above, we conclude that:

$$(\widetilde{X}_{1j}^T\widetilde{X}_{1j} - \widetilde{X}_{1j}^T\widetilde{P}_{mj}\widetilde{X}_{1j}) = \widetilde{X}_{1j}^T\widetilde{X}_{1j} - \widetilde{X}_{1j}^T\widetilde{P}_{mj}\widetilde{X}_{1j} = \frac{1}{a}(X_{1j}^T X_{1j} - X_{1j}^T P_{mj}X_{1j})$$

$$(I_m - \widetilde{P}_{mj})\widetilde{X}_{1j} = \frac{1}{\sqrt{a}}(I_m - P_{mj})(I_m - \delta J_m)X_{1j} = \frac{1}{\sqrt{a}}(I_m - P_{mj})X_{1j}$$

$$(\widetilde{X}_1^T\widetilde{X}_1 - \widetilde{X}_1^T\widetilde{P}\widetilde{X}_1) = \sum_{j=0}^{q}(\widetilde{X}_{1j}^T\widetilde{X}_{1j} - \widetilde{X}_{1j}^T\widetilde{P}_{mj}\widetilde{X}_{1j}) = \frac{1}{a}(X_1^T X_1 - X_1^T P X_1)$$

Corollary 1. *The designs that result in the estimators with the minimum variance, i.e., $\text{minvar}(\hat{\tau})$ are exactly the optimal designs of the model with independent observations. In this case, the variance $\text{var}(\hat{\tau})$ is multiplied by α:*

$$\text{var}(\tau) = \sigma^2(X_1^T X_1 - X_1^T P X_1)^{-1} = \sigma^2 a \cdot (Q^*)^{-1}$$

$\sigma^2(Q^*)^{-1}$ is the variance of the optimal designs in the model with independent observations).

Proof. From the previous proof, we conclude that the variance of the estimator of the direct effect, which is given by Formula (3), equals to

$$\text{var}(\tau) = \sigma^2 a \cdot (Q^*)^{-1}$$

□

Comments: (1) If we consider that an observation can influence another observation, the e.u are correlated, and the correlation is given by ρ, $-1 < \rho < 1$. Dependent observations are often considered observations of the same cluster [17]. A simple example of dependency appears when children of the same mother are included in a sample. Due to their common household environment and genes, it is expected that these children have a greater chance of having the same characteristics.

(2) In the case of compound symmetry, the variance matrix of each sequence observations is $\Sigma_m = (1-\rho)\mathbf{I}_m + \rho\mathbf{J}_m$, so $\alpha = 1-\rho$, and $b = \rho$. In order for the matrix to be positive definite, the condition $-\frac{1}{m-1} < \rho < 1$ is necessary. If $\rho = 0$, then we obtain the model with independent observations and $\alpha = 1$.

(3) The variance of the estimator of the direct effect, $\text{var}(\hat{\tau})$, decreases when the correlation coefficient ρ increases and it approaches 0, when ρ approaches 1, since $\alpha = 1-\rho$.

(4) For two periods with dependent observations, the 2×2 variance matrix of the observations in the compound symmetry model is $\Sigma_2 = (1-\rho)\mathbf{I}_2 + \rho\mathbf{J}_2$. The optimal design for this model is the same as the optimal design for independent observations for every ρ, $-1 < \rho < 1$.

For an even n, such an optimal design is obtained when to the sequences AA and AB correspond to $n/2$ e.u, while for an odd n, the optimal design is obtained when to the sequences AA and AB correspond to $(n-1)/2$ and $(n+1)/2$ e.u., respectively [11]. The reverse sequences BB, BA also correspond to an optimal design with: $\text{var}(\tau) = \sigma^2(1-\rho)(\mathbf{Q}^*)^{-1}$.

(5) As illustrated, the examined model with dependent observations is also associated with variance matrices Σ for which the optimal designs are the same as the ones of the model with independent observations [14,18].

Funding: This research received no external funding.

Conflicts of Interest: The author declares no conflict of interest.

References

1. Carriere, K.C.; Reinsel, G.C. Optimal two period repeated measurement designs with two or more treatments. *Biometrika* **1993**, *80*, 924–929. [CrossRef]
2. Chalikias, M.; Kounias, S. Extension and necessity of Cheng and Wu conditions. *J. Stat. Plan. Infer.* **2012**, *142*, 1794–1800. [CrossRef]
3. Cheng, C.S.; Wu, C.F. Balanced repeated measurements designs. *Ann. Stat.* **1980**, *11*, 29–50. [CrossRef]
4. Hedayat, A.S.; Yang, M. Universal Optimality of Selected Crossover Designs. *J. Am. Stat. Assoc.* **2004**, *99*, 461–466. [CrossRef]
5. Hedayat, A.S.; Zheng, W. Optimal and efficient crossover designs for test-control study when subject effects are random. *J. Am. Stat. Assoc.* **2010**, *105*, 1581–1592. [CrossRef]
6. Stufken, J. Some families of optimal and efficient repeated measurements designs. *J. Stat. Plan. Infer.* **1991**, *27*, 75–83. [CrossRef]
7. Hedayat, A.S.; Yan, Z. Crossover designs based on type I orthogonal arrays for a self and simple mixed carryover effects model with correlated errors. *J. Stat. Plan. Infer.* **2008**, *138*, 2201–2213. [CrossRef]
8. Kounias, S.; Chalikias, M.S. An algorithm applied to designs of repeated measurements. *J. Appl. Stat. Sci.* **2005**, *14*, 243–250.
9. Kushner, H.B. Allocation rules for adaptive repeated measurements designs. *J. Stat. Plan. Infer.* **2003**, *113*, 293–313. [CrossRef]
10. Laska, E.M.; Meisner, M. A variational approach to optimal two treatment crossover designs: Application to carryover effect models. *J. Am. Stat. Assoc.* **1985**, *80*, 704–710. [CrossRef]
11. Matthews, J.N.S. Optimal crossover designs for the comparison of two treatments in the presence of carryover effects and autocorrelated errors. *Biometrika* **1987**, *74*, 311–320. [CrossRef]
12. Hedayat, A.; Afsarinejad, K. Repeated measurements designs, I. *Survey Stat. Des. Linear Models* **1975**, 229–242. Available online: http://ani.stat.fsu.edu/techreports/M261.pdf (accessed on 23 April 2019).
13. Hedayat, A.S.; Afsarinejad, K. Repeated measurements designs II. *Ann. Stat.* **1978**, *18*, 1805–1816. [CrossRef]

14. Kounias, S.; Chalikias, M. *Optimal and Universally Optimal Two Treatment Repeated Measurement Designs*; Vonta, F., Nikulin, M., Eds.; Statistics for industry and technology Birkhauser: Boston, MA, USA; Basel, Switzerland; Berlin, Germany, 2008; pp. 465–477.
15. Kounias, S.; Chalikias, M.S. Optimal two treatment repeated measurement designs with treatment-period interaction in the model. *Util. Math.* **2015**, *96*, 243–261.
16. Kounias, S.; Chalikias, M. Estimability of Parameters in a Linear Model. *Stat. Probab. Lett.* **2008**, *28*, 2437–2439. [CrossRef]
17. Liang, K.Y.; Zeger, S.L. Regression analysis for correlated data. *Annu. Rev. Pub. Health* **1993**, *14*, 43–68. [CrossRef] [PubMed]
18. Chalikias, M.; Kounias, S. Optimal two Treatment Repeated Measurement Designs for three Periods. *Commun. Stat. Theory Methods* **2017**, *46*, 200–209. [CrossRef]

© 2019 by the author. Licensee MDPI, Basel, Switzerland. This article is an open access article distributed under the terms and conditions of the Creative Commons Attribution (CC BY) license (http://creativecommons.org/licenses/by/4.0/).

Article
A Model for Predicting Statement Mutation Scores

Lili Tan, Yunzhan Gong and Yawen Wang *

State Key Laboratory of Networking and Switching Technology, Beijing University of Posts and Telecommunications, Beijing 100876, China
* Correspondence: wangyawen@bupt.edu.cn

Received: 13 June 2019; Accepted: 15 August 2019; Published: 23 August 2019

Abstract: A test suite plays a key role in software testing. Mutation testing is a powerful approach to measure the fault-detection ability of a test suite. The mutation testing process requires a large number of mutants to be generated and executed. Hence, mutation testing is also computationally expensive. To solve this problem, predictive mutation testing builds a classification model to predict the test result of each mutant. However, the existing predictive mutation testing methods only can be used to estimate the overall mutation scores of object-oriented programs. To overcome the shortcomings of the existing methods, we propose a new method to directly predict the mutation score for each statement in process-oriented programs. Compared with the existing predictive mutation testing methods, our method uses more dynamic program execution features, which more adequately reflect dynamic dependency relationships among the statements and more accurately reflects information propagation during the execution of test cases. By comparing the prediction effects of logistic regression, artificial neural network, random forest, support vector machine, and symbolic regression, we finally decide to use a single hidden layer feedforward neural network as the predictive model to predict the statement mutation scores. In our two experiments, the mean absolute errors between the statement mutation scores predicted by the neural network and the real statement mutation scores both approximately reach 0.12.

Keywords: software testing; machine learning; mutation testing

1. Introduction

When a programmer writes a program, a mistake may occur in the code. For example, a programmer may incorrectly write x=x-1 as x=x+1, x=x*1, x=x%1, etc. This mistake is referred to as a software fault (i.e., a software bug). When this fault is executed, an incorrect execution result may appear on the corresponding statement. This incorrect execution result often is referred to as a software error and cannot be directly observed. When this software error propagates to an observable program output, a software failure occurs.

A strong-power test suite may detect more software faults than a weak-power one, thus measuring the fault detection capability of a test suite is an important question in software testing. Mutation testing is an approach to determine the effectiveness of a test suite [1–3].

The programs with software faults are called mutants. In mutation testing, mutants are generated through automatically changing the original program with mutation operators, where each mutation operator is a rule and can be applied to program statements to produce the program version with a software fault. A mutant is said to be identified by a test suite if at least one test case from the test suite has different execution results on the mutant and the original program. Mutation score, which is the ratio of all identified mutants to all mutants, has been widely used to assess the adequacy of a test suite.

Although mutation testing is obviously useful, it is extremely expensive [4,5]. For example, using 108 mutation operators, Proteum [6] generates 4937 mutants for tcas, which is the smallest program among the Siemens programs and contains only 137 non-commenting and non-whitespace lines of code. Thus, testing a large number of mutants can be a big burden.

For solving this problem, researchers have proposed some optimization methods to reduce the cost of mutation testing, such as random mutation [7,8], mutant clustering [9] and selective mutation [10,11]. For quickly calculating the mutation score of the whole program, these methods attempt to use a mutant sample to represent all mutants. Random mutation randomly chooses some mutants from all mutants to construct mutation samples. A mutant clustering algorithm first classifies all mutants into different clusters so that the mutants in a cluster have similar identification difficulties, and then selects a small number of mutants from each cluster to construct the mutant sample. Selective mutation uses only a subset of mutation operators to generate a mutant sample.

Different from the above mutant reduction methods, the predictive mutation testing methods [12,13] have been proposed in recent years. The predictive mutation testing methods extract some features related to program structures and testing processes and apply machine learning to predict each mutant's test result (i.e., the identification result). Moreover, these predictive methods' execution time is short. However, the existing predictive mutation testing methods are all designed for object-oriented programs. The same as other methods, the existing predictive mutation testing methods are also mainly used for estimating the mutation score of the whole program. The main differences among the above mutant reduction methods can be shown in Table 1.

Table 1. Main differences among mutation reduction methods.

Method	Key Technology	Time Cost	Target
random mutation	simple random sampling	low	estimating program mutation score
mutant clustering	stratified sampling	low	estimating program mutation score
selective mutation	non-probability sampling	high	estimating program mutation score
predictive mutation	supervised learning	low	estimating program mutation score classifying mutants

To make up for the shortcomings of existing predictive mutation testing methods, based on the execution impact map [14] Goradia uses, we suggest a new predictive method. This new method is not only suitable for procedure-oriented programs but also can use a single hidden layer feedforward neural network and seven statement features to predict the mutation score of each program statement.

The prediction of the statement mutation scores includes two major phases: extracting the statement features and determining the mathematical form of predictive model. In the feature extraction phase, we obtain the following seven features to express the effect of a statement on the program outputs: number of executions, path impact factor, value impact factor, generalized path impact factor, generalized value impact factor, latent impact factor, and information hidden factor. In fact, among the above seven features, only a number of executions are adopted by existing predictive mutation testing methods. Compared with the existing predictive mutation testing methods, our method more accurately expresses information propagation among the statements. For a statement, except for the number of executions, its six other features are extracted from the following six aspects respectively:

When a test case executes on the statement containing a software fault, an error may be generated. This error either propagates along the original execution path or changes the original execution path.

(1) The fault in the statement may change the program output by generating the errors that propagate along the original execution paths. From this aspect, we extract the statement's value impact factor.

(2) The fault in the statement may change the program outputs by generating the errors altering the original execution paths. From this aspect, we extract the statement's path impact factor.

However, in a few cases, the change of execution path does not result in a change of program output. Therefore, we need to analyze further the features of the changed program branch in order to more accurately predict how likely the program output will be changed.

(3) The no longer executed branches lose their ability to pass their information along the original execution path to the program outputs. The loss of this capability may cause the program output to be changed. From this aspect, we extract the statement's generalized value impact factor.

(4) The no longer executed program branch is no longer able to influence the selection of subsequent program branches. Loss of this ability may also impact the program output. From this aspect, we extract the statement's generalized path impact factor.

(5) The fault in a statement may cause some program branches, which has not been executed, will be executed. Executing these new branches may cause the program output to change. From this aspect, we extract the statement's latent impact factor.

(6) Sometimes, the program under testing has multiple output statements, some of which happen to have the same output values. In this case, even if the software fault changes the execution path of the test case, the program outputs could still be the same. From this aspect, we extract a statement's information hidden factor.

Among these six factors, the first five factors facilitate program output changes, and the last one prevents program output from changing.

In the phase of determining mathematical form of the predictive model, we compared the following five machine learning models based on Brier scores: artificial neural network (ANN), logical regression (LR), random forest (RF), support vector machine (SVM) and symbolic regression (SR). From the experiment results, the artificial neural networks were identified as the most suitable predictive model.

With the methods in this article, we analyzed the two programs. In the two experiments, the mean absolute errors between the real statement mutation scores and predictive statement mutation scores are 0.1205 and 0.1198, respectively.

The remainder of this paper is organized as below: in Section 2, we introduce some basic terms used throughout the entire paper. In Section 3, we define seven statement features. In Section 4, we propose a method for quickly calculating statement features. In Section 5, we compare the prediction accuracy of five machine learning models. In Section 6, we introduce the structure of our automated prediction tool. In Section 7, we describe the work to be performed.

2. Basic Terms

Definition 1. *Original program and mutation score.*

In this paper, a program without any software fault is also called an original program. For example, Program 1 is an original program. It first outputs the factorial of the absolute value of the difference between m and n, and then classifies the factorial. Based on the relationships among m, n and the factorial, the execution results of the program are divided into three areas, the first and third of which belong to the first class, and the second of which belongs to the second class.

A program with software faults is called a mutant. In mutation testing, mutants are generated through automatically changing the original program with mutation operators. For example, in terms of Program 1, if the statement dist=m-n is changed into dist=m%n, then the mutant m_1 is generated as shown in Program 2. If a test suite (i.e., a collection of test cases) can identify the mutant m_1, it must satisfy the following conditions: there must be at least one test case in the test suite to execute the statement dist=m%n in m_1, the execution result of dist=m%n must be different from that of dist=m-n, and the difference must be propagated to the program output.

Program mutation score is the proportion of identified mutants in a program, which is used to assess how well the program is tested by the test suite. Statement mutation score is the the proportion of identified mutants in a statement, which is used to assess how well the statement is tested by the test suite.

Definition 2. *Program statement and branch.*

In this article, we predict the ability of a test suite to test each line program code. A statement in the program under testing usually occupies one line. Because a control expression usually occupies a line in the program, in this paper, we also think of a controlling expression as a statement. As shown in Program 1, we denote gth statement as s_g. According to C programming language standard—C99 [15], a controlling expression can occur in "if", "switch", "while", "do while" and "for" statements and decides which of the program branches is executed.

In terms of if-else statement, if its controlling expression appears in the rth line, then we denoted its controlling expression as s_r, and use $B_{r,t}$ and $B_{r,f}$ to denote the true branch and false branch of s_r, respectively. In terms of a loop statement (such as while loop, do-while loop and for loop), we regard it as the combination of the controlling expression and the corresponding program branch. If a loop statement's controlling expression appears in the rth line, then its controlling expression is denoted as s_r, and the corresponding loop body is considered as the true branch of s_r, so that this loop body can also be denoted as $B_{r,t}$. According to this representation method, the program branch whose function is to exit the loop is denoted as $B_{r,f}$.

Program 1: An original program.

	#include <stdio.h>
	typedef int bool ;
	void fun(int m, int n) {
	int dist, fac ;
s_1	if(m>n)
s_2	dist=m-n ;
	else
s_3	dist=n-m;
s_4	fac=1;
s_5	while (dist>1) { // Loop for factorial
s_6	fac=fac ∗ dist ;
s_7	dist = dist -1 ;
	}
s_8	printf ("fac=%d \n", fac);
s_9	if (m<n) // classify the factorial
s_{10}	printf ("class 1 \n") ;
s_{11}	else if (fac<5)
s_{12}	printf ("class 2 \n") ;
	else
s_{13}	printf ("class 1 \n") ;
	}

Program 2: The mutant m_1 of Program 1.

	#include <stdio.h>
	typedef int bool ;
	void fun(int m, int n) {
	int dist, fac ;
s_1	if(m>n)
s_2	dist=m%n ;
	else
s_3	dist=n-m;
s_4	fac=1;
s_5	while (dist>1) { // Loop for factorial
s_6	fac=fac ∗ dist ;
s_7	dist = dist -1 ;
	}
s_8	printf ("fac=%d \n", fac);
s_9	if (m<n) // classify the factorial
s_{10}	printf ("class 1 \n") ;
s_{11}	else if (fac<5)
s_{12}	printf ("class 2 \n") ;
	else
s_{13}	printf ("class 1 \n") ;
	}

For example, in Program 1, s_9 is the controlling expression, the statement s_{10} constitutes its true branch $B_{9,t}$, and the statements s_{11}, s_{12} and s_{13} constitute its false branch $B_{9,f}$. The statements s_6 and s_7 constitute the loop body of the while loop, and, in this situation, the loop body is also considered as the true branch $B_{5,t}$ of the controlling expression s_5.

Definition 3. *Statement instance and branch instance.*

A statement may be executed multiple times by a test suite, so that multiple execution instances are generated. The statement's each execution instance is called its a statement instance. The hth execution instance of test case t_k on statement s_g is denoted as s_{g,t_k}^h. In this paper, the execution instance of a program output statement is called an *output statement instance*. In addition, the execution instance of a controlling expression is also considered as a special statement instance, and is called a *controlling expression instance*.

For example, when Program 1 is executed by test case $t_1(m=4, n=1)$, the assignment statement s_4, controlling expression s_5, controlling expression s_9, controlling expression s_{11} and output statement s_{13} are executed once, three times, once, once and once. This allows them to produce one, three, one, one and one execution instance, respectively, during the execution of the test case t_1. Among them, the controlling expression instances s_{5,t_1}^1, s_{5,t_1}^2 and s_{5,t_1}^3, respectively, represent the first, second and third executions of the test case t_1 on the statement s_5.

A program branch may also be executed multiple times, so that many execution instances are generated. Each execution instance of the program branch is called *a branch instance*. Just as a program branch consists of many statements, a branch instance consists of many statement instances. These statement instances are called the *statement instances in the branch instance*. A bit similar to the symbols of statement instances, we use B_{r,z,t_k}^l to represent the lth execution instance of the test case t_k on the program branch $B_{r,z}$, where z represents the true or the false branch, and its value is t or f. Whether $B_{r,z}$ is executed depends on the execution result of the controlling expression s_r.

For example, the branch instance B_{9,t,t_1}^1 consists of s_{10,t_1}^1, and the branch instance B_{9,f,t_1}^1 consists of s_{11,t_1}^1, s_{12,t_1}^1 and s_{13,t_1}^1. In terms of the while statement in Program 1, s_5 is a controlling expression and generates three execution instances s_{5,t_1}^1, s_{5,t_1}^2 and s_{5,t_1}^3 during the execution of the test case t_1. Because the execution of B_{5,t,t_1}^1 is the necessary condition for B_{5,t,t_1}^2 to be executed, B_{5,t,t_1}^2 is contained in B_{5,t,t_1}^1. As shown in Table 2, Figures 1 and 2, the first branch instance B_{5,t,t_1}^1 of the while loop consists of the statement instances s_{6,t_1}^1, s_{7,t_1}^1, s_{5,t_1}^2, s_{6,t_1}^2, s_{7,t_1}^2 and s_{5,t_1}^3, and the second branch instance B_{5,t,t_1}^2 consists of the statement instances s_{6,t_1}^2, s_{7,t_1}^2 and s_{5,t_1}^3.

Table 2. The execution history of the test cases.

Test Case	Program Output	Execution History	Branch Instances in Loop
m = 4, n = 1	fac = 6, class 1 (s_{13})	$H_1 : s_{1,t_1}^1, s_{2,t_1}^1, s_{4,t_1}^1, s_{5,t_1}^1, s_{6,t_1}^1, s_{7,t_1}^1, s_{5,t_1}^2,$ $s_{6,t_1}^2, s_{7,t_1}^2, s_{5,t_1}^3, s_{8,t_1}^1, s_{9,t_1}^1, s_{11,t_1}^1, s_{13,t_1}^1$	$B_{5,t,t_1}^1 = \{s_{6,t_1}^1, s_{7,t_1}^1, s_{5,t_1}^2, s_{6,t_1}^2, s_{7,t_1}^2, s_{5,t_1}^3\}, B_{5,t,t_1}^2 = \{s_{6,t_1}^2, s_{7,t_1}^2, s_{5,t_1}^3\}$
m = 2, n = 2	fac = 1, class 2 (s_{12})	$H_2 : s_{1,t_2}^1, s_{3,t_2}^1, s_{4,t_2}^1, s_{5,t_2}^1, s_{8,t_2}^1, s_{9,t_2}^1, s_{11,t_2}^1, s_{12,t_2}^1$	
m = 1, n = 4	fac = 6, class 1 (s_{10})	$H_3 : s_{1,t_3}^1, s_{3,t_3}^1, s_{4,t_3}^1, s_{5,t_3}^1, s_{6,t_3}^1, s_{7,t_3}^1, s_{5,t_3}^2,$ $s_{6,t_3}^2, s_{7,t_3}^2, s_{5,t_3}^3, s_{8,t_3}^1, s_{9,t_3}^1, s_{10,t_3}^1$	$B_{5,t,t_3}^1 = \{s_{6,t_3}^1, s_{7,t_3}^1, s_{5,t_3}^2, s_{6,t_3}^2, s_{7,t_3}^2, s_{5,t_3}^3\}, B_{5,t,t_3}^2 = \{s_{6,t_3}^2, s_{7,t_3}^2, s_{5,t_3}^3\}$

Definition 4. *Original execution path of the test case.*

The execution history H_k of the test case t_k is formed when the test case t_k executes on an original program. The execution history H_k is an execution trace, each element of which is a statement instance. These statement instances are ordered by time until the last program output. In this paper, the execution history H_k of the test case t_k is also called the original execution path of t_k.

For example, consider the Program 1, where test case t_1 (m = 4, n = 1), test case t_2 (m = 2, n = 2), and test case t_3 (m = 1, n = 4) constitutes the test suite T. As shown in Table 2, when t_1 is executed, H_1 is generated, and the program outputs fac = 6 and class 1. When t_2 is executed, H_2 is generated, and the program outputs fac = 1 and class 2. When t_3 is executed, H_3 is generated and the program outputs fac = 6 and class 1.

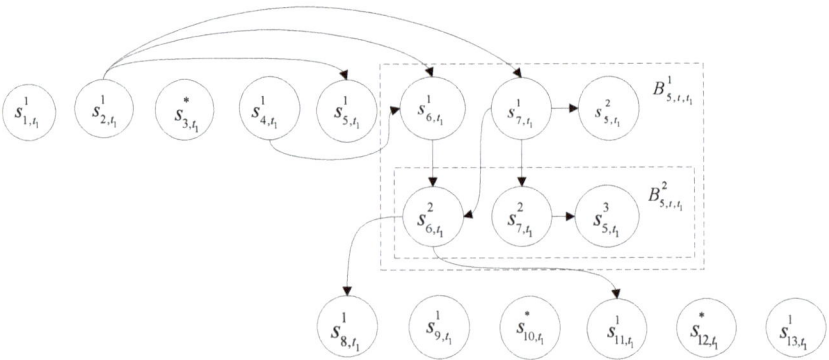

Figure 1. The execution impact graph G_1 formed when Program 1 is executed by test case 1.

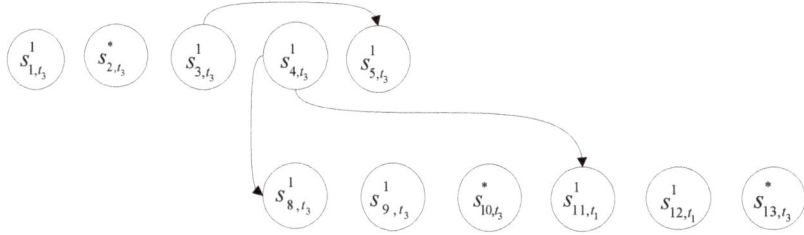

Figure 2. The execution impact graph G_2 formed when Program 1 is executed by test case 2.

Definition 5. *Execution impact graph*

An execution impact graph G_k is formed when the test case t_k executes. The execution impact graph G_k consists of multiple impact arcs generally, and each impact arc expresses the information propagation between the statement instances. In terms of an impact arc, the arc tail s_{i,t_k}^j is called a direct impact predecessor, and the arc head s_{g,t_k}^h is called a direct impact successor. In the practical application, if a variable is assigned in the statement instance s_{i,t_k}^j and is directly used at the statement instance s_{g,t_k}^h, then s_{i,t_k}^j is a direct impact predecessor of s_{g,t_k}^h, and s_{g,t_k}^h is a direct impact successor of s_{i,t_k}^j. In the execution impact graph G_k, each node is expressed in the form of s_{i,t_k}^j or s_{i,t_k}^*, where s_{i,t_k}^j denotes a statement instance and the symbol $*$ indicates that the statement s_i is not executed by test case t_k.

For example, when program 1 is executed by test cases 1, 2, and 3, the corresponding execution impact graphs are generated respectively, as shown in Figures 1, 2, and 3. In Program 1, the variable dist is defined in the statement s_2 and is directly used in the statements s_5, s_6 and s_7. Hence, when the test case t_1 is executed, s_{2,t_1}^1 becomes the direct impact predecessor of s_{5,t_1}^1, s_{6,t_1}^1 and s_{7,t_1}^1, respectively. In this situation, s_{5,t_1}^1, s_{6,t_1}^1 and s_{7,t_1}^1 become the direct impact successors of s_{2,t_1}^1.

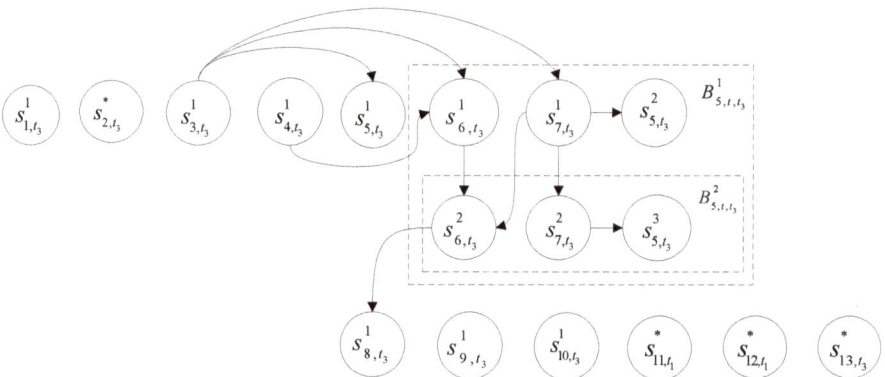

Figure 3. The execution impact graph G_3 formed when Program 1 is executed by test case 3.

Each direct impact successor of a statement instance may have its own direct impact successor. Thus, the impact successor is transitive. If a statement instance is the impact successor of the statement instance s_{g,t_k}^h but it is not the direct impact successor of s_{g,t_k}^h, then this statement instance is called the indirect impact successor of s_{g,t_k}^h. Thus, the impact successor can be divided into two types: the direct impact successor and the indirect successor.

47

For example, $s^1_{5,t_1}, s^1_{6,t_1}, s^1_{7,t_1}, s^2_{5,t_1}, s^2_{6,t_1}, s^2_{7,t_1}, s^3_{5,t_1}, s^1_{8,t_1}$ and s^1_{11,t_1} are all the impact successors of s^1_{2,t_1}. However, s^1_{5,t_1}, s^1_{6,t_1} and s^1_{7,t_1} are the direct impact successors of s^1_{2,t_1}, and $s^2_{5,t_1}, s^2_{6,t_1}, s^2_{7,t_1}, s^3_{5,t_1}, s^1_{8,t_1}$ and s^1_{11,t_1} are the indirect impact successors of s^1_{2,t_1}.

If there is a fault f_g in statement s_g, and s^h_{g,t_k} is an execution instance of statement s_g, then f_g may change the execution result of s^h_{g,t_k} during the execution of the test case t_k. If this change happens, we say that a error e^h_{g,t_k} is generated from the statement instance s^h_{g,t_k}. In this paper, an error is different from a fault. Errors are dynamic and are generated in the process of the test case execution. However, faults are static. Whether the program under testing is executed or not, they may exist in the program under testing.

3. Formal Definitions of Statement Features

In this section, we propose the seven features of a statement. The most of them are related to execution paths of test cases. When the statement containing a software fault is executed by a test case, an error may generate. After this error generates, it either propagates along the original execution path of the test case or changes the original execution path. The value impact factor describes the ability of the fault existing in a statement to affect the program output under the condition that the execution path is unchanged. The path impact factor, the generalized value impact factor, the generalized path impact factor and the latent impact factor describe the abilities of the fault existing in a statement to affect the program output under the condition that the execution path is changed by the generated error.

3.1. Value Impact Factor

The value impact factor of a statement expresses its ability to directly impact the program outputs along the execution paths of the test cases.

3.1.1. Value Impact Factor of Statement

The errors generated from the statement instance s^h_{g,t_k} may propagate along the original execution path H_k to some execution instances of the output statements. Each of these output statement instances is called the *value impact element of the statement instance* s^h_{g,t_k}. The collection consisting of all value impact elements of s^h_{g,t_k} is called *value impact set of the statement instance* s^h_{g,t_k}, and denoted as V^h_{g,t_k}.

A statement s_g has multiple execution instances generally and each execution instance has its own value impact set. The union of these value impact sets is called the *value impact set of* s_g, and is denoted as V_g. The element in V_g is called the value impact element of s_g. The number of value impact elements of s_g is called the value impact factor of s_g, and is denoted as $x_{vi}(s_g)$. Therefore, the following formula holds:

$$V_g = \bigcup_{k=1,2,\cdots,K} \bigcup_{h=1,2,\cdots,H_{gk}} V^h_{g,t_k}, \qquad (1)$$

where K is the total number of test cases in the test suite, and H_{gk} is the total number of times the statement s_g is executed by the test case t_k.

Example 1. *From Table 2, we know that the statement s_6 has four execution instances $s^1_{6,t_1}, s^2_{6,t_1}, s^1_{6,t_3}$ and s^2_{6,t_3}. If s_6 includes a fault, then each execution instance of s_6 may generate an error. The errors generated from s^1_{6,t_1} and s^2_{6,t_1} may propagate along the original execution path H_1 to the output statement instance s^1_{8,t_1}. Therefore, $V^1_{6,t_1} = V^2_{6,t_1} = \{s^1_{8,t_1}\}$. The errors generated from s^1_{6,t_3} and s^2_{6,t_3} may propagate along the original execution path H_3 to the output statement instance s^1_{8,t_3}. Therefore, $V^1_{6,t_3} = V^2_{6,t_3} = \{s^1_{8,t_3}\}$. According to Formula (1), we have $V_6 = V^1_{6,t_1} \cup V^2_{6,t_1} \cup V^1_{6,t_3} \cup V^2_{6,t_3} = \{s^1_{8,t_1}, s^1_{8,t_3}\}$.*

3.1.2. The Value Impact Relationship between Statement Instance and Its Direct Impact Successors

According to the relationship between the impact precursor and the impact successor, we get the following conclusion: If the statement instances $s^{q_1}_{p_1,t_k}, s^{q_2}_{p_2,t_k}, \cdots, s^{q_n}_{p_n,t_k}$ are all direct impact successors of the statement instance s^h_{g,t_k}, then we get

$$V^h_{g,t_k} = \bigcup_{c=1,2,\cdots,n} V^{q_c}_{p_c,t_k}. \qquad (2)$$

Example 2. *From Figure 1, we can know that the direct impact successors of the statement instance s^1_{7,t_1} consist of s^2_{5,t_1}, s^2_{6,t_1} and s^2_{7,t_1}. Under the condition that we know $V^2_{5,t_1} = \varnothing$, $V^2_{6,t_1} = \{s^1_{8,t_1}\}$ and $V^2_{7,t_1} = \varnothing$, we have*

$$V^1_{7,t_1} = V^2_{5,t_1} \cup V^2_{6,t_1} \cup V^2_{7,t_1} = \{s^1_{8,t_1}\}.$$

This formula indicates that the errors generated from the statement instance s^1_{7,t_1} can reach up to one output statement instance s^1_{8,t_1} when it propagates along the original execution path of the test case t_1. Using the same method, we also know $V^1_{6,t_1} = \{s^1_{8,t_1}\}$, $V^2_{6,t_1} = \{s^1_{8,t_1}\}$, $V^2_{5,t_1} = \varnothing$ and $V^3_{5,t_1} = \varnothing$.

3.1.3. Value Impact Set of Branch Instance

The information expressed by the statement instances in the branch instance B^l_{r,z,t_k} can propagate along the original execution path H_k to some execution instances of the program output statements. These affected output statement instances constitute the *value impact set* V^l_{r,z,t_k} of the branch instance B^l_{r,z,t_k}. We can get the following formula:

$$V^l_{r,z,t_k} = \bigcup_{d=1,2,\cdots,n} V^{h_d}_{g_d,t_k}, \qquad (3)$$

where $s^{h_1}_{g_1,t_k}, s^{h_2}_{g_2,t_k}, \cdots, s^{h_n}_{g_n,t_k}$ are all the statement instances in the branch instance B^l_{r,z,t_k}.

Example 3. *We can use formula (3) to calculate the value impact set of the branch instance B^1_{5,t,t_1}. From Example 2, we know both $V^1_{6,t_1} = \{s^1_{8,t_1}\}$, $V^1_{7,t_1} = \{s^1_{8,t_1}\}$, $V^2_{5,t_1} = \varnothing$, $V^2_{6,t_1} = \{s^1_{8,t_1}\}$ $V^2_{7,t_1} = \varnothing$, $V^3_{5,t_1} = \varnothing$. Because the branch instance B^1_{5,t,t_1} consists of the six statement instances $s^1_{6,t_1}, s^1_{7,t_1}, s^2_{5,t_1}, s^2_{6,t_1}, s^2_{7,t_1},$ and s^3_{5,t_1}, we get*

$$V^1_{5,t,t_1} = V^1_{6,t_1} \cup V^1_{7,t_1} \cup V^2_{5,t_1} \cup V^2_{6,t_1} \cup V^2_{7,t_1} \cup V^3_{5,t_1} = \{s^1_{8,t_1}\}.$$

3.1.4. Value Impact Set of the Special Statement Instance

If a statement instance is an output statement instance, it usually does not have any impact successors. We set its value impact set to itself because the change of its execution result is precisely the change of program output. If a statement instance is not an output statement instance and does not have any impact successors, then we set its value impact set to an empty set.

3.2. Path Impact Factor

The path impact factor of a statement expresses its ability to directly impact the execution paths of the test cases.

3.2.1. Path Impact Factor of Statement

The more controlling expression instances a statement impact, the more easily the fault in the statement changes the execution paths of the test cases. The more likely the execution path is changed, the more likely the program output will be changed. Therefore, we take the number of the control expression instances impacted by a statement during the test suite execution as a feature to describe the effect of this statement on program output. For this purpose, we defined a statement's path impact factor.

The errors generated from the statement instance s_{g,t_k}^h may propagate along the original execution path H_k to some controlling expression instances. The collection of these controlling expression instances is called the *path impact set* P_{g,t_k}^h of the statement instance s_{g,t_k}^h. The element in P_{g,t_k}^h is called the *path impact element of* s_{g,t_k}^h. The path impact set of statement s_g is the union of path impact sets of execution instances of s_g, and denoted as P_g. In other words,

$$P_g = \bigcup_{k=1,2,\cdots,K} \bigcup_{h=1,2,\cdots,H_{gk}} P_{g,t_k}^h. \tag{4}$$

K is the total number of test cases in the test suite, and H_{gk} is the total number of times the statement s_g is executed by the test case t_k.

Example 4. *From Table 2, we know that the statement s_6 has four execution instances s_{6,t_1}^1, s_{6,t_1}^2, s_{6,t_3}^1 and s_{6,t_3}^2. If s_6 includes a fault, then when s_6 is executed by the test suite, each execution instance may generate an error. The errors generated from first two statement instances s_{6,t_1}^1 and s_{6,t_1}^2 may propagate along the original execution path H_1 to the controlling expression instance s_{11,t_1}^1. Along the original execution path H_3, the errors generated from the last two statement instances s_{6,t_3}^1 and s_{6,t_3}^2 cannot be propagated to any controlling expression instance. Therefore, $P_{6,t_1}^1 = P_{6,t_1}^2 = \{s_{11,t_1}^1\}$ and $P_{6,t_3}^1 = P_{6,t_3}^2 = \emptyset$. Using Formula (4), we get*

$$P_6 = P_{6,t_1}^1 \bigcup P_{6,t_1}^2 \bigcup P_{6,t_3}^1 \bigcup P_{6,t_3}^2 = \{s_{11,t_1}^1\}.$$

3.2.2. The Path Impact Relationship of the Statement Instance and Its Direct Impact Successor

According to the relationship between the impact precursor and the impact successor, we get the following conclusion: If the statement instances $s_{p_1,t_k}^{q_1}, s_{p_2,t_k}^{q_2}, \cdots, s_{p_n,t_k}^{q_n}$ are all direct impact successors of the statement instance s_{g,t_k}^h, then

$$P_{g,t_k}^h = \bigcup_{c=1,2,\cdots,n} P_{p_c,t_k}^{q_c}. \tag{5}$$

Example 5. *From Figure 1, we can know the the direct impact successors of the statement instance s_{7,t_1}^1 consist of s_{5,t_1}^2, s_{6,t_1}^1 and s_{7,t_1}^2. Under the condition that we know $P_{5,t_1}^2 = \{s_{5,t_1}^2\}$, $P_{6,t_1}^1 = \{s_{11,t_1}^1\}$ and $P_{7,t_1}^2 = \{s_{5,t_1}^3\}$, according to Formula (5), we have*

$$P_{7,t_1}^1 = P_{5,t_1}^2 \bigcup P_{6,t_1}^1 \bigcup P_{7,t_1}^2 = \{s_{5,t_1}^2, s_{11,t_1}^1, s_{5,t_1}^3\}.$$

Therefore, the errors generated from the statement instance s_{7,t_1}^1 can change up to three controlling expression instances s_{5,t_1}^2, s_{11,t_1}^1 and s_{5,t_1}^3 along the original execution path H_1. Using the same method, we can also get $P_{6,t_1}^1 = \{s_{11,t_1}^1\}$, $P_{5,t_1}^3 = \{s_{5,t_1}^3\}$, and so on.

3.2.3. Path Impact Set of Branch Instance

The statement instances in the branch instance B_{r,z,t_k}^l may propagate their information along the original execution path H_k to some of the controlling expression instances outside of B_{r,z,t_k}^l. These controlling expression instances constitute the *path impact set* P_{r,z,t_k}^l of the branch instance B_{r,z,t_k}^l. The path impact set of B_{r,z,t_k}^l express the impact of B_{r,z,t_k}^l on the controlling expression instances outside of B_{r,z,t_k}^l. If $s_{g_1,t_k}^{h_1}, s_{g_2,t_k}^{h_2}, \cdots, s_{g_n,t_k}^{h_n}$ are all the statement instances in the branch instance B_{r,z,t_k}^l, then the following mathematical formula holds

$$P_{r,z,t_k}^l = \left(\bigcup_{d=1,2,\cdots,n} P_{g_d,t_k}^{h_d} \right) \setminus B_{r,z,t_k}^l. \tag{6}$$

Example 6. We illustrate the formula above by calculating the path impact set of the branch instance B^1_{5,t,t_1}. From Example 5, we know that $P^1_{6,t_1} = \{s^1_{11,t_1}\}$, $P^1_{7,t_1} = \{s^2_{5,t_1}, s^3_{5,t_1}, s^1_{11,t_1}\}$, $P^2_{5,t_1} = \{s^2_{5,t_1}\}$, $P^2_{6,t_1} = \{s^1_{11,t_1}\}$, $P^2_{7,t_1} = \{s^3_{5,t_1}\}$, and $P^3_{5,t_1} = \{s^3_{5,t_1}\}$. Because the branch instance B^1_{5,t,t_1} consists of the six statement instances $s^1_{6,t_1}, s^1_{7,t_1}, s^2_{5,t_1}, s^2_{6,t_1}, s^2_{7,t_1}$ and s^3_{5,t_1}, we get

$$P^1_{5,t,t_1} = (P^1_{6,t_1} \cup P^1_{7,t_1} \cup P^2_{5,t_1} \cup P^2_{6,t_1} \cup P^2_{7,t_1} \cup P^3_{5,t_1}) \setminus B^1_{5,t,t_1} = \{s^1_{11,t_1}\}.$$

3.2.4. Path Impact Set of the Special Statement Instance

If a statement instance is a controlling expression instance, it usually does not have any impact successors. We set its path impact set to itself because the change of its execution result is precisely the change of the program execution path. If a statement instance is not a controlling expression instance and does not have any impact successors, then we set its path impact set to an empty set.

3.3. Generalized Value Impact Factor

The generalized value impact factor of a statement expresses its ability to indirectly impact the program outputs.

3.3.1. Generalized Value Impact Factor of Statement

The error generated from the statement instance s^h_{g,t_k} may propagate to some controlling expression instances along the original execution path of test case t_k, so that the execution results of these controlling expression instances may be changed. As long as the execution result of the control expression instance s^l_{r,t_k} is changed, the branch instance B^l_{r,z,t_k}, which appears in the original execution path H_k, will no longer be executed. This makes the statement instances in B^l_{r,z,t_k} no longer pass their information to some output statement instances. Thus, the execution results of these output statement instances may be changed. Therefore, the errors generated from the statement instance s^h_{g,t_k} may indirectly affect some output statement instances through the above error propagation process. These output statement instances that may be indirectly influenced by s^h_{g,t_k} form the *generalized value impact set* of the statement instance s^h_{g,t_k}. The generalized value impact set of the statement instance s^h_{g,t_k} is denoted as $\mathcal{V}^h_{g,t_k}$. The element in $\mathcal{V}^h_{g,t_k}$ is called the generalized value impact element of s^h_{g,t_k}. The number of generalized value impact element of s^h_{g,t_k} is called the generalized value impact factor of s^h_{g,t_k}, and is denoted as $x_{gvi}(s^h_{g,t_k})$.

A statement s_g has multiple execution instances generally and each execution instance has its own generalized value impact set. In order to describe this indirect effect of s_g on program output, the union of these generalized value impact sets is called the generalized value impact set of s_g. The generalized value impact set of s_g is denoted as $\mathcal{V}_g$, the element in $\mathcal{V}_g$ is called the generalized value impact element of s_g, and the number of the generalized value impact element of s_g is called the generalized value impact factor of s_g. In summary,

$$\mathcal{V}_g = \bigcup_{k=1,2,\cdots,K} \bigcup_{h=1,2,\cdots,H_{gk}} \mathcal{V}^h_{g,t_k}, \qquad (7)$$

where K is the total number of test cases in the test suite, and H_{gk} is the total number of times the statement s_g is executed by the test case t_k.

Example 7. In Program 1, the statement s_7 has four execution instances $s^1_{7,t_1}, s^2_{7,t_1}, s^1_{7,t_3}$, and s^2_{7,t_3}. Given that $\mathcal{V}^1_{7,t_1} = \{s^1_{8,t_1}, s^1_{13,t_1}\}$, $\mathcal{V}^2_{7,t_1} = \emptyset$, $\mathcal{V}^1_{7,t_3} = \{s^1_{8,t_3}\}$, and $\mathcal{V}^2_{7,t_3} = \emptyset$, we can use Formula (7) to calculate the generalized value impact set of the statement s_7:

$$\mathcal{V}_7 = \mathcal{V}^1_{7,t_1} \cup \mathcal{V}^2_{7,t_1} \cup \mathcal{V}^1_{7,t_3} \cup \mathcal{V}^2_{7,t_3} = \{s^1_{8,t_1}, s^1_{13,t_1}, s^1_{8,t_3}\}.$$

3.3.2. Generalized Value Impact Set of the Special Statement Instance

A controlling expression instance s^l_{r,t_k} usually does not have any impact successors. Corresponding to s^l_{r,t_k}, there is usually a branch instance B^l_{r,z,t_k} that appears in the original execution path H_k. In this situation, the generalized value impact set of s^l_{r,t_k} is equal to the value impact set of the branch instance B^l_{r,z,t_k}. This conclusion can be interpreted as follows: If an error is generated from the controlling expression instance s^l_{r,t_k}, then the branch instance B^l_{r,z,t_k} will no longer be executed, so that the statement instances in B^l_{r,z,t_k} can no longer propagate their information along the original execution path H_k to some output statement instances. This error propagation process also exactly reflects the impact of branch instance B^l_{r,z,t_k} on program output. Hence, the above conclusion is proved. For example, the generalized value impact set of the controlling expression instance s^1_{5,t_1} is equal to the value impact set of the branch instance B^1_{5,t,t_1}.

If a statement instance is not a controlling expression instance and does not have any impact successors, then we set its generalized value impact set to an empty set.

3.3.3. The Generalized Value Impact Relationship between a Statement and Its Direct Impact Successors

According to the relationship between the impact precursor and the impact successor, we get the following conclusion: If the statement instances $s^{q_1}_{p_1,t_k}, s^{q_2}_{p_2,t_k}, \cdots, s^{q_n}_{p_n,t_k}$ are all direct impact successors of the statement instance s^h_{g,t_k}, then

$$V^h_{g,t_k} = \bigcup_{c=1,2,\cdots,n} V^{q_c}_{p_c,t_k}. \tag{8}$$

Example 8. *In Program 1, s^2_{5,t_1}, s^2_{6,t_1} and s^2_{7,t_1} are all direct impact successors of the statement instance s^1_{7,t_1}. Given that $V^2_{5,t_1} = \{s^1_{8,t_1}\}$, $V^2_{6,t_1} = \{s^1_{13,t_1}\}$, and $V^2_{7,t_1} = \emptyset$, according to formula (8), we can get*

$$V^1_{7,t_1} = V^2_{5,t_1} \bigcup V^2_{6,t_1} \bigcup V^2_{7,t_1} = \{s^1_{8,t_1}, s^1_{13,t_1}\}.$$

In the same way, given that $V^3_{5,t_1} = \emptyset$, we can get $V^2_{7,t_1} = V^3_{5,t_1} = \emptyset$. Given that $V^2_{5,t_3} = \{s^1_{8,t_3}\}$, $V^2_{6,t_3} = \emptyset$, and $V^2_{7,t_3} = \emptyset$, we can get $V^1_{7,t_3} = V^2_{5,t_3} \bigcup V^2_{6,t_3} \bigcup V^2_{7,t_3} = \{s^1_{8,t_3}\}$

3.4. Generalized Path Impact Factor

The generalized path impact factor of a statement expresses its ability to indirectly change the program execution path.

3.4.1. Generalized Path Impact Factor of Statement

The error generated from the statement instance s^h_{g,t_k} may propagate to some controlling expression instances along the original execution path of test case t_k. As long as the execution result of the control expression instance s^l_{r,t_k} is changed, the branch instance B^l_{r,z,t_k} that appears in the original execution path H_k will no longer be executed. The statement instances in B^l_{r,z,t_k} will no longer pass their information to the controlling expression instances appearing after B^l_{r,z,t_k}. In this situation, the execution results of the controlling expression instances appearing after B^l_{r,z,t_k} may be changed because they are no longer influenced by the statement instances in B^l_{r,z,t_k}. Therefore, the errors generated from the statement instance s^h_{g,t_k} may indirectly affect some controlling expression instances appearing after B^l_{r,z,t_k} through the above error propagation process. These controlling expression instances that may be indirectly affected by s^h_{g,t_k} through the above error propagation process form the generalized path impact set of the statement instance s^h_{g,t_k}. This set is denoted as $\mathcal{P}^h_{g,t_k}$, the element of which is called the generalized path impact element of s^h_{g,t_k}. The number of generalized path impact elements of s^h_{g,t_k} is called the generalized path impact factor of s^h_{g,t_k}, and is denoted as $x_{gpi}(s^h_{g,t_k})$.

A statement s_g has one or more execution instances generally. Therefore, the generalized path impact set of s_g is defined as the union of generalized path impact sets of the execution instances of s_g, and is denoted as $\mathcal{P}_g$. In other words,

$$\mathcal{P}_g = \bigcup_{k=1,2,\cdots,K} \bigcup_{h=1,2,\cdots,H_{gk}} \mathcal{P}_{g,t_k}^h, \tag{9}$$

where K is the total number of test cases in the test suite, and H_{gk} is the total number of times the statement s_g is executed by the test case t_k. The element in $\mathcal{P}_g$ is called the generalized path impact element of s_g. The number of generalized path impact element of s_g is called the generalized path impact factor of s_g, and denoted as $x_{gpi}(s_g)$.

Example 9. *We explain the above definitions by calculating the generalized path impact set of the statement s_7. In Program 1, the statement s_7 has four execution instances s_{7,t_1}^1, s_{7,t_1}^2, s_{7,t_3}^1 and s_{7,t_3}^2. Given that $\mathcal{P}_{7,t_1}^1 = \{s_{11,t_1}^1\}$, $\mathcal{P}_{7,t_1}^2 = \varnothing$, $\mathcal{P}_{7,t_3}^1 = \varnothing$ and $\mathcal{P}_{7,t_3}^2 = \varnothing$, we can use Formula (9) to calculate the generalized value impact set of the statement s_7.*

$$\mathcal{P}_7 = \mathcal{P}_{7,t_1}^1 \bigcup \mathcal{P}_{7,t_1}^2 \bigcup \mathcal{P}_{7,t_3}^1 \bigcup \mathcal{P}_{7,t_3}^2 = \{s_{11,t_1}^1\}.$$

3.4.2. Generalized Path Impact Set of the Special Statement Instance

If a statement instance s_{r,t_k}^l is a controlling expression instance, then it usually does not have any impact successors. Corresponding to s_{r,t_k}^l, there is usually a branch instance B_{r,z,t_k}^l, which exists in the original execution path H_k. In this situation, the generalized path impact set of s_{r,t_k}^l is precisely the path impact set of B_{r,z,t_k}^l. This conclusion can be interpreted as follows: Assume there is a software fault in statement s_r. If an error is generated from the controlling expression instance s_{r,t_k}^l, then the branch instance B_{r,z,t_k}^l will no longer be executed, the information expressed by the statement instances in B_{r,z,t_k}^l can no longer propagate along the original execution path H_k to some controlling expression instances outside of B_{r,z,t_k}^l. This error propagation process also exactly reflects the impact of branch instance B_{r,z,t_k}^l on the execution path of the test case t_k. Therefore, the generalized path impact set of s_{r,t_k}^l is equal to the path impact set of B_{r,z,t_k}^l. For example, the generalized path impact set of the controlling expression instance s_{5,t_1}^1 is equal to the path impact set of the branch instance B_{5,t,t_1}^1. In other words, $\mathcal{P}_{5,t_1}^1 = P_{5,t,t_1}^1 = \{s_{11,t_1}^1\}$. Otherwise, if a statement instance is not a controlling expression instance and does not have any impact successors, then we set its generalized path impact set to an empty set.

3.4.3. The Generalized Path Impact Relationship between a Statement Instance and Its Direct Impact Successors

According to the relationship between the impact precursor and the impact successor, we get the following conclusion: If the statement instances $s_{p_1,t_k}^{q_1}$, $s_{p_2,t_k}^{q_2}$, $\cdots$, $s_{p_n,t_k}^{q_n}$ are all direct impact successors of the statement instance s_{g,t_k}^h, then

$$\mathcal{P}_{g,t_k}^h = \bigcup_{c=1,2,\cdots,n} \mathcal{P}_{p_c,t_k}^{q_c}. \tag{10}$$

Example 10. *In Program 1, s_{5,t_1}^2, s_{6,t_1}^2 and s_{7,t_1}^2 are all direct impact successors of the statement instance s_{7,t_1}^1. Given that $\mathcal{P}_{5,t_1}^2 = \{s_{11,t_1}^1\}$, $\mathcal{P}_{6,t_1}^2 = \varnothing$, and $\mathcal{P}_{7,t_1}^2 = \varnothing$, according to formula (10), we can get*

$$\mathcal{P}_{7,t_1}^1 = \mathcal{P}_{5,t_1}^2 \bigcup \mathcal{P}_{6,t_1}^2 \bigcup \mathcal{P}_{7,t_1}^2 = \{s_{11,t_1}^1\}.$$

3.5. Latent Impact Factor

The fault in a statement may cause some program branches that have not yet been executed to be executed. The latent impact factor expresses the impact of these branches to be executed on the program output.

3.5.1. Latent Impact Factor of the Program Statement

Contrary to the branch instances that will no longer be executed, some branch instances may be going to be executed due to the error generated from the statement instance s_{g,t_k}^h. These branches to be executed may change the program outputs. For an example, in Program 1, if the assignment statement s_2 is mutated into dist=m%n, then the remainder dist becomes zero when test case t_1 runs. In this situation, the true branch $B_{11,t}$ of s_{11}, which consists of s_{12} and does not appear in the original execution path H_1, will be executed and change the program output.

These branch instances to be executed are divided into two classes. In the first class, each branch instance contains statement instances. In the second class, each branch instance does not. The first class branch instances constitute the *latent impact set of statement instance* s_{g,t_k}^h, and denotes as L_{g,t_k}^h. The element in L_{g,t_k}^h is called the *latent impact element of the statement instance* s_{g,t_k}^h. The number of latent impact elements of s_{g,t_k}^h is called the *latent impact factor of* s_{g,t_k}^h and denoted as $x_{li}(s_{g,t_k}^h)$.

A statement s_g has multiple execution instances generally, and each of them has its own latent impact set. Therefore, the union of these latent impact sets is defined as the *latent impact set of the statement* s_g, and denoted as L_g. In other words,

$$L_g = \bigcup_{k=1,2,\cdots,K} \bigcup_{h=1,2,\cdots,H_{gk}} L_{g,t_k}^h, \tag{11}$$

where K is the total number of test cases in the test suite, and H_{gk} is the total number of times the statement s_g is executed by the test case t_k. The element in L_g is called the *latent impact element of* s_g. The number of latent impact element of s_g is called the *latent impact factor of the statement* s_g, and denoted as $x_{li}(s_g)$.

Example 11. *We are going to calculate the latent impact factor of the statement s_7. As shown in Table 2, s_7 has the four execution instances s_{7,t_1}^1, s_{7,t_1}^2, s_{7,t_3}^1 and s_{7,t_3}^2. Assume four errors e_{7,t_1}^1, e_{7,t_1}^2, e_{7,t_3}^1 and e_{7,t_3}^2 are generated from the statement instances s_{7,t_1}^1, s_{7,t_1}^2, s_{7,t_3}^1 and s_{7,t_3}^2, respectively. In this situation, e_{7,t_1}^1 may propagate along the original execution path H_1 to the controlling expression instances s_{5,t_1}^2, s_{5,t_1}^3 and s_{11,t_1}^1. When e_{7,t_1}^1 propagates to s_{5,t_1}^2, the branch instances B_{5,f,t_1}^2 that do not appear in the original execution path H_1 will be executed. However, the role of $B_{5,f}$ is to exit the loop, so that it does not contain any statements. Thus, B_{5,f,t_1}^2 itself does not affect program output. This makes B_{5,f,t_1}^2 not a latent impact element of s_{7,t_1}^1. When e_{7,t_1}^1 propagates along the original execution path H_1 to s_{5,t_1}^3, the branch instance B_{5,t,t_1}^3 that does not appear in the original execution path H_1 will be executed. The $B_{5,t}$ contains some statements so that the execution of B_{5,t,t_1}^3 in itself may change the program outputs. Thus, B_{5,t,t_1}^3 is a latent impact element of s_{7,t_1}^1. When e_{7,t_1}^1 propagates along the original execution path H_1 to s_{11,t_1}^1, the branch instance B_{11,t,t_1}^1 that does not appear in the original execution path H_1 will be executed. The program branch $B_{11,t}$ contains some statements so that the execution of B_{11,t,t_1}^1 in itself may change the program outputs. Thus, the branch instance B_{11,t,t_1}^1 is a latent impact element of s_{7,t_1}^1. From the above analysis, we can know that the latent impact set of s_{7,t_1}^1 consists of B_{5,t,t_1}^3 and B_{11,t,t_1}^1. In the similar way, we can know that the latent impact set of the statement instance s_{7,t_1}^2 consists of B_{5,t,t_1}^3. The latent impact set of s_{7,t_3}^1 consists of B_{5,t,t_3}^3, and that of s_{7,t_3}^2 also consists of B_{5,t,t_3}^3. With Formula (11), we can get the latent impact set of the statement s_7:*

$$L_7 = L_{7,t_1}^1 \bigcup L_{7,t_1}^2 \bigcup L_{7,t_3}^1 \bigcup L_{7,t_3}^2 = \{B_{5,t,t_1}^3, B_{11,t,t_1}^1, B_{5,t,t_3}^3\}.$$

3.5.2. The Latent Impact Relationship between a Statement Instance and its Direct Impact Successors

According to the relationship between the impact precursor and the impact successor, we get the following conclusion: If the statement instances $s_{p_1,t_k}^{q_1}, s_{p_2,t_k}^{q_2}, \cdots, s_{p_n,t_k}^{q_n}$ are all direct impact successors of the statement instance s_{g,t_k}^h, then

$$L_{g,t_k}^h = \bigcup_{c=1,2,\cdots,n} L_{p_c,t_k}^{q_c}. \tag{12}$$

Example 12. *From Figure 1, we can know the the direct impact successors of the statement instance s_{7,t_1}^1 consist of s_{5,t_1}^2, s_{6,t_1}^2 and s_{7,t_1}^2. Given that $L_{5,t_1}^2 = \varnothing$, $L_{6,t_1}^2 = \{B_{11,t,t_1}^1\}$ and $L_{7,t_1}^2 = \{B_{5,t,t_1}^3\}$, according to Formula (12), we have*

$$L_{7,t_1}^1 = L_{5,t_1}^2 \bigcup L_{6,t_1}^2 \bigcup L_{7,t_1}^2 = \{B_{11,t,t_1}^1, B_{5,t,t_1}^3\}.$$

In addition, under the condition that we know $L_{5,t_3}^2 = \varnothing$, $L_{6,t_3}^2 = \varnothing$ and $L_{7,t_3}^2 = \{B_{5,t,t_3}^3\}$, using the same method, we can still get

$$L_{7,t_3}^1 = L_{5,t_3}^2 \bigcup L_{6,t_3}^2 \bigcup L_{7,t_3}^2 = \{B_{5,t,t_3}^3\}.$$

Furthermore, we can get

$$L_7 = L_{7,t_1}^1 \bigcup L_{7,t_1}^2 \bigcup L_{7,t_3}^1 \bigcup L_{7,t_3}^2 = \{B_{11,t,t_1}^1, B_{5,t,t_1}^3, B_{5,t,t_3}^3\}.$$

3.5.3. Latent Impact Set of the Special Statement Instance

If a statement instance s_{r,t_k}^l is a controlling expression instance and the branch instance B_{r,z,t_k}^l does not appear in the original execution path H_k; then, in the condition that B_{r,z,t_k}^l is not empty, we set B_{r,z,t_k}^l as the only element in the latent impact set of s_{r,t_k}^l. If a statement instance is not a controlling expression instance and does not have any impact successors, then we set the latent impact set of s_{r,t_k}^l to an empty set.

For example, as far as the controlling expression instance s_{5,t_1}^2 is concerned, although the branch instance B_{5,f,t_1}^2 does not appear in the original execution path H_1, B_{5,f,t_1}^2 does not include any statement instance. Hence, B_{5,f,t_1}^2 is not a latent impact element of s_{5,t_1}^2, and we set the latent impact set of s_{5,t_1}^2 to an empty set. As far as the controlling expression instance s_{5,t_1}^3 is concerned, because the branch instance B_{5,t,t_1}^3 not only does not appear in the original execution path H_1 but also is not empty, we set s_{5,t_1}^3 as the only element in the latent impact set of B_{5,t,t_1}^3.

3.6. Information Hidden Factor

The last feature of a statement is its information hiding feature. Sometimes, the program has multiple output statements, and some of them happen to generate same outputs. In this case, even if the software fault in a statement changes the execution path of the test case, the output of the program may still not be changed.

This phenomenon make the faults in statements difficult to identify. For a statement s_g, we use the information hiding factor to express this feature. The information hiding factor of s_g can be calculated in the following way. We use the test cases that execute s_g to construct sub test suite T_g. When we execute T_g, the program under testing generates some outputs. The information entropy of the output distribute is called the information hidden factor of statement s_g, and denoted as $x_{ih}(s_g)$. In other words,

$$x_{ih}(s_g) = -\sum_i p_i \log_2 p_i, \tag{13}$$

where p_i is the probability that the test cases executing the statement s_g generate the ith program output class.

Example 13. *We calculate the information hidden factors of the statements s_9 and s_{11}, respectively. In Program 1, the test suite consists of the test cases t_1, t_2 and t_3. These three test cases all execute statement s_9. Their executions generate three program outputs (fac = 6, class 1), (fac = 1 class 2) and (fac = 6 class 1), respectively. Hence, the probability that the program output (fac = 6, class 1) is 0.67, and the probability that the program output (fac = 1 class 2) is 0.33. According to Formula (13), the information hidden factor of statement s_9 is 0.9182 bit. The test cases t_1 and t_2 execute the statement s_{11}. Their executions generate two program outputs (fac = 6, class 1) and (fac = 1 class 2), respectively. Hence, the probabilities that the program output (fac = 6, class 1) and (fac = 6, class 1) are both 0.5. According to Formula (13), the information hidden factor of statement s_{11} is 1.0 bit.*

4. Calculation of Statement Features

First, we propose an iterative method to compute statement features, and then compare the time cost of this method with that of direct mutant testing.

4.1. Calculation Process

We divide the calculation of all the statement features into two parts. The first part calculation includes the first five statement features: the value impact factor, the path impact factor, the generalized value impact factor, the generalized path impact factor, and the latent impact factor. The second part calculation includes includes the last two statement features: the number of times a statement is executed, and information hidden factor.

The first part of the calculation takes much more time than the second one. For reducing the computational complexity, we propose an iterative method. Generally, if a statement instance has at least one impact successor, then we can calculate its first five features according to the formulas (2), (5), (8), (10), and (12). Otherwise, we use the methods mentioned in Sections 3.1.4, 3.2.4, 3.3.2, 3.4.2 and 3.5.3 to calculate its first five features.

The computation of the statement features is divided into two corresponding stages. The first stage, including steps 1–6, calculate the first part of statement features. The second stage including steps 7 and 8, calculate the second part of statement features. The overall computation steps are as follows:

Step 1 Set test case serial number $k = 1$.

Step 2 Construct the execution impact graph G_k of the test case t_k.

Step 3 First, from the original execution path of the test case t_k, find all statement instances that have not been analyzed. From these unanalyzed statement instances, find the last executed statement instance. We might as well denote this statement instance as s_{g,t_k}^h.

(1) If s_{g,t_k}^h has one or more impact successors, then we construct the impact sets of its first five features according to the formulas (2), (5), (8), (10) and (12).

(2) If s_{g,t_k}^h does not have any impact successors, then we construct the impact sets of its first five features according to the methods mentioned in Sections 3.1.4, 3.2.4, 3.3.2, 3.4.2 and 3.5.3.

Step 4 If there are some statement instances which appear in the original execution path of test case t_k but have not yet been analyzed, go to step 3, else go to step 5.

Step 5 If test case t_k is not the last test case in test suite, then k = k + 1, and go to step 2, else go to step 6.

Step 6 First, construct each program statement's value impact set, path impact set, generalized value impact set, generalized path impact set and latent impact set by formulas (1), (4), (7), (9) and (11). Next, for each program statement, calculate its value impact factor, path impact factor, generalized value impact factor, generalized path impact factor factor and the latent impact factor.

Step 7 For each statement in program under testing, compute the total number of times it is executed by the test cases in the test suite.

Step 8 For each statement in the program under testing, compute its information hidden factor by formula (13).

Example 14. *We illustrate the above process by extracting the features of each statement in Program 1. In terms of the first stage of extracting the statement features, whether the statement instances are generated during the execution of test case 1, test case 2 or test case 3, the methods for calculating features of the statement instance are the same. Therefore, with regard to steps 1 to 5, we only explain in detail how to calculate the features of the statement instances generated during test case t_1 execution. The detailed calculation process is as follows.*

We first set $k = 1$, execute test case t_1, and construct the execution impact graph G_1 of test case t_1 as shown in Figure 1.

The first analyzed statement instance is the last executed statement instance in original execution path H_1. Thus, we first analyze the output statement instance s_{13,t_1}^1. According to Sections 3.1.4, 3.2.4, 3.3.2, 3.4.2 and 3.5.3, we get $V_{13,t_1}^1 = \{s_{13,t_1}^1\}$, $P_{13,t_1}^1 = \varnothing$, $\mathcal{V}_{13,t_1}^1 = \varnothing$, $\mathcal{P}_{13,t_1}^1 = \varnothing$ and $Ł_{13,t_1}^1 = \varnothing$.

The second analyzed statement instance s_{11,t_1}^1 is the penultimate element in original execution path H_1. Because it is a controlling expression, we get $V_{11,t_1}^1 = \varnothing$, $P_{11,t_1}^1 = \{s_{11,t_1}^1\}$, $\mathcal{V}_{11,t_1}^1 = V_{11,t,t_1}^1 = V_{13,t_1}^1 = \{s_{13,t_1}^1\}$, $\mathcal{P}_{11,t_1}^1 = \varnothing$ and $L_{11,t_1}^1 = \{B_{11,t,t_1}^1\}$ according to Sections 3.1.4, 3.2.4, 3.3.2, 3.4.2 and 3.5.3.

The third analyzed statement instance s_{9,t_1}^1 is the antepenultimate element in H_1. Because s_{9,t_1}^1 is a controlling expression instance, we get $V_{9,t_1}^1 = \varnothing$, $P_{9,t_1}^1 = \{s_{9,t_1}^1\}$, $\mathcal{V}_{9,t_1}^1 = V_{9,t,t_1}^1 = V_{11,t_1}^1 \cup V_{13,t_1}^1 = \{s_{13,t_1}^1\}$, $\mathcal{P}_{9,t_1}^1 = P_{9,t,t_1}^1 = (P_{11,t_1}^1 \cup P_{13,t_1}) \setminus B_{9,t_1}^1 = (s_{11,t_1}^1 \cup \varnothing) \setminus B_{9,t,t_1}^1 = \varnothing$ and $Ł_{9,t_1}^1 = \{B_{9,t,t_1}^1\}$ according to Sections 3.1.4, 3.2.4, 3.3.2, 3.4.2 and 3.5.3.

The fourth analyzed statement instant s_{8,t_1}^1 is the fourth element from the end of H_1. Because s_{8,t_1}^1 is an output statement instance, $V_{8,t_1}^1 = \{s_{8,t_1}^1\}$, $P_{8,t_1}^1 = \varnothing$, $\mathcal{V}_{8,t_1}^1 = \varnothing$, $\mathcal{P}_{8,t_1}^1 = \varnothing$ and $Ł_{8,t_1}^1 = \varnothing$ according to Sections 3.1.4, 3.2.4, 3.3.2, 3.4.2 and 3.5.3.

The fifth analyzed statement instant s_{5,t_1}^3 is the fifth element from the end of H_1. Because s_{5,t_1}^3 is a controlling expression instance of zero length, $V_{5,t_1}^3 = \varnothing$, $P_{5,t_1}^3 = \{s_{5,t_1}^3\}$, $\mathcal{V}_{5,t_1}^3 = V_{5,f,t_1}^3 = \varnothing$, $\mathcal{P}_{5,t_1}^3 = P_{5,t,t_1}^1 = \varnothing$ and $Ł_{5,t_1}^3 = \{B_{5,t,t_1}^3\}$ according to Sections 3.1.4, 3.2.4, 3.3.2, 3.4.2 and 3.5.3.

The sixth analyzed statement instance s_{7,t_1}^2 is the sixth statement instance from the end of H_1. Because the direct impact successors of s_{7,t_1}^2 consist of s_{5,t_1}^3, we get $V_{7,t_1}^2 = V_{5,t_1}^3 = \varnothing$, $P_{7,t_1}^2 = P_{5,t_1}^3 = \{s_{5,t_1}^3\}$, $\mathcal{V}_{7,t_1}^2 = \mathcal{V}_{5,t_1}^3 = \varnothing$, $\mathcal{P}_{7,t_1}^2 = \mathcal{P}_{5,t_1}^3 = \varnothing$, $Ł_{7,t_1}^2 = Ł_{5,t_1}^3 = \{B_{5,t,t_1}^3\}$ according to formulas (2), (5), (8), (10) and (12).

The seventh analyzed statement instance s_{6,t_1}^2 is the seventh element from the end of H_1. Becasue the direct impact successors of s_{6,t_1}^2 consist of s_{8,t_1}^1 and s_{11,t_1}^1, we get $V_{6,t_1}^2 = V_{8,t_1}^1 \cup V_{11,t_1}^1 = \{s_{8,t_1}^1\}$, $P_{6,t_1}^2 = P_{8,t_1}^1 \cup P_{11,t_1}^1 = \{s_{11,t_1}^1\}$, $\mathcal{V}_{6,t_1}^2 = \mathcal{V}_{8,t_1}^1 \cup \mathcal{V}_{11,t_1}^1 = \{s_{13,t_1}^1\}$, $\mathcal{P}_{6,t_1}^2 = \mathcal{P}_{8,t_1}^1 \cup \mathcal{P}_{11,t_1}^1 = \varnothing$ and $L_{6,t_1}^2 = Ł_{8,t_1}^1 \cup Ł_{11,t_1}^1 = \{B_{11,t,t_1}^1\}$ according to formulas (2), (5), (8), (10) and (12).

The eighth analyzed statement instance s_{5,t_1}^2 is the eighth element from the end of H_1. Because s_{5,t_1}^2 is a controlling expression instance, we get $V_{5,t_1}^2 = \varnothing$, $P_{5,t_1}^2 = \{s_{5,t_1}^2\}$, $\mathcal{V}_{5,t_1}^2 = V_{5,t,t_1}^2 = V_{6,t_1}^2 \cup V_{7,t_1}^2 \cup V_{5,t_1}^3 = \{s_{8,t_1}^1\}$, $\mathcal{P}_{5,t_1}^2 = P_{5,t,t_1}^2 = (P_{6,t_1}^2 \cup P_{7,t_1}^2 \cup P_{5,t_1}^3) \setminus P_{5,t,t_1}^2 = \{s_{11,t_1}^1\}$ and $Ł_{5,t_1}^2 = \varnothing$ according to Sections 3.1.4, 3.2.4, 3.3.2, 3.4.2 and 3.5.3.

The ninth analyzed statement instance s_{7,t_1}^1 is the ninth element from the end of H_1. Because the direct impact successors of s_{7,t_1}^1 consist of s_{5,t_1}^2, s_{6,t_1}^2 and s_{7,t_1}^2, we get $V_{7,t_1}^1 = V_{5,t_1}^2 \cup V_{6,t_1}^2 \cup V_{7,t_1}^2 = \{s_{8,t_1}^1\}$, $P_{7,t_1}^1 = P_{5,t_1}^2 \cup P_{6,t_1}^2 \cup P_{7,t_1}^2 = \{s_{5,t_1}^2, s_{5,t_1}^3, s_{11,t_1}^1\}$, $\mathcal{V}_{7,t_1}^1 = \mathcal{V}_{5,t_1}^2 \cup \mathcal{V}_{6,t_1}^2 \cup \mathcal{V}_{7,t_1}^2 = \{s_{8,t_1}^1, s_{13,t_1}^1\}$, $\mathcal{P}_{7,t_1}^1 = \mathcal{P}_{5,t_1}^2 \cup \mathcal{P}_{6,t_1}^2 \cup \mathcal{P}_{7,t_1}^2 = \{s_{11,t_1}^1\}$ and $L_{7,t_1}^1 = Ł_{5,t_1}^2 \cup Ł_{6,t_1}^2 \cup Ł_{7,t_1}^2 = \{B_{5,t,t_1}^3, B_{11,t,t_1}^1\}$ according to formulas (2), (5), (8), (10) and (12).

The tenth analyzed statement instance s_{6,t_1}^1 is the tenth element from the end of H_1. Because the direct impact successors of s_{6,t_1}^1 consist of s_{6,t_1}^2, we get $V_{6,t_1}^1 = V_{6,t_1}^2 = \{s_{8,t_1}^1\}$, $P_{6,t_1}^1 = P_{6,t_1}^2 = \{s_{11,t_1}^1\}$,

$\mathcal{V}^1_{6,t_1} = \mathcal{V}^2_{6,t_1} = \{s^1_{13,t_1}\}$, $\mathcal{P}^1_{6,t_1} = \mathcal{P}^2_{6,t_1} = \emptyset$ and $L^1_{6,t_1} = \{B_{11,t,t_1}\}$ according to formulas (2), (5), (8), (10) and (12).

The eleventh analyzed statement instance s^1_{5,t_1} is the eleventh statement instance from the end from H_1. Because s^1_{5,t_1} is a controlling expression instance, we get $V^1_{5,t_1} = \emptyset$, $P^1_{5,t_1} = \{s^1_{5,t_1}\}$, $\mathcal{V}^1_{5,t_1} = \mathcal{V}^1_{5,t,t_1} = \{s^1_{8,t_1}\}$, $\mathcal{P}^1_{5,t_1} = \mathcal{P}^1_{5,t,t_1} = \{s^1_{11,t_1}\}$ and $L^1_{5,t_1} = \emptyset$ according to Sections 3.1.4, 3.2.4, 3.3.2, 3.4.2 and 3.5.3.

The twelfth analyzed statement instance s^1_{4,t_1} is the twelfth element from the end of H_1. Because the direct impact successors of s^1_{4,t_1} consist of s^1_{6,t_1}, we get $V^1_{4,t_1} = V^1_{6,t_1} = \{s^1_{8,t_1}\}$, $P^1_{4,t_1} = P^1_{6,t_1} = \{s^1_{11,t_1}\}$, $\mathcal{V}^1_{4,t_1} = \mathcal{V}^1_{6,t,t_1} = \{s^1_{13,t_1}\}$, $\mathcal{P}^1_{4,t_1} = \mathcal{P}^1_{6,t,t_1} = \emptyset$, $L^1_{4,t_1} = L^1_{6,t_1} = \{B^1_{11,t,t_1}\}$ according to formulas (2), (5), (8), (10) and (12).

The thirteenth analyzed statement instance s^1_{2,t_1} is the thirteenth element from the end of H_1. The direct impact successors of s^1_{2,t_1} consist of s^1_{5,t_1}, s^1_{6,t_1} and s^1_{7,t_1}. According to formulas (2), (5), (8), (10) and (12), we get $V^1_{2,t_1} = V^1_{5,t_1} \cup V^1_{6,t_1} \cup V^1_{7,t_1} = \{s^1_{8,t_1}\}$, $P^1_{2,t_1} = P^1_{5,t_1} \cup P^1_{6,t_1} \cup P^1_{7,t_1} = \{s^1_{5,t_1}, s^2_{5,t_1}, s^3_{5,t_1}, s^1_{11,t_1}\}$, $\mathcal{V}^1_{2,t_1} = \mathcal{V}^1_{5,t_1} \cup \mathcal{V}^1_{6,t_1} \cup \mathcal{V}^1_{7,t_1} = \{s^1_{8,t_1}, s^1_{13,t_1}\}$, $\mathcal{P}^1_{2,t_1} = \mathcal{P}^1_{5,t_1} \cup \mathcal{P}^1_{6,t_1} \cup \mathcal{P}^1_{7,t_1} = \{s^1_{11,t_1}\}$ and $L^1_{2,t_1} = L^1_{5,t_1} \cup L^1_{6,t_1} \cup L^1_{7,t_1} = \{B^3_{5,t,t_1}, B^1_{11,t,t_1}\}$.

The fourteenth analyzed statement instance s^1_{1,t_1} is the fourteenth element from the end of H_1. Because s^1_{1,t_1} is a controlling expression instance, we get $V^1_{1,t_1} = \emptyset$, $P^1_{1,t_1} = \{s^1_{1,t_1}\}$, $\mathcal{V}^1_{1,t_1} = \mathcal{V}^1_{1,t,t_1} = \mathcal{V}^1_{2,t_1} = \{s^1_{8,t_1}\}$, $\mathcal{P}^1_{1,t_1} = \mathcal{P}^1_{1,t,t_1} = \mathcal{P}^1_{2,t_1} \setminus B^1_{1,t,t_1} = \{s^1_{5,t_1}, s^2_{5,t_1}, s^3_{5,t_1}, s^1_{11,t_1}\}$ and $L^1_{1,t_1} = B^1_{1,f,t_1}$ according to Sections 3.1.4, 3.2.4, 3.3.2, 3.4.2 and 3.5.3.

Similar to the above procedure, we can continue to calculate the above first five impact sets for all statement instances generated during the executions of test case 2 and test case 3, respectively. Thus far, steps 1–5 are completed, and their final results are shown in Table 3.

After calculating the first five impact sets of each statement instance, based on Table 3 and Formulas (1), (4), (7), (9) and (11), we can get the first five impact sets of each statement, as shown in Table 4. The corresponding impact factors are shown in Table 5.

After calculating the first five impact factors of each statement, we calculate the execution number of each statement, as shown in Table 6.

Finally, similar to Example 13, we use the Formula (13) to calculate the information hidden factor of each statement. The calculation process are shown in Table 7.

4.2. Computational Complexity Analysis

Through the analysis of computational complexity, we can draw the following conclusion: Compared with the time required for direct mutation testing, the time used to calculate all statement features can be neglected. The computation time of statement features consists of two parts. The first part of time overhead is used to calculate the value impact factor, path impact factor, generalized value impact factor, generalized path impact factor and potential impact factor. The calculation of these features is relatively complex. The second part of time overhead is used to calculate the other two statement features. The calculation of these two features is relatively simple. Therefore, we can approximatively consider the first part of time overhead as the total time overhead for computing all statement features. In this situation, to prove the conclusion, we only need to prove that the first part of time overhead is much lower than that used to directly execute mutation testing. We can get this conclusion from the following four steps:

In the first steps, we first suppose the time overhead that the computer spends to executes one statement once is T_0. According the Section 4.1, we can get the following conclusion: the time overhead used to compute a factor of a statement instance is also roughly equal to T_0. If a statement instance has at least one impact successor, then we use the formulas (2), (5), (8), (10) and (12) to calculate its first five impact factors, respectively. Otherwise, this statement does not have any impact successors, and we use the method in the Sections 3.1.4, 3.2.4, 3.3.2, 3.4.2 and 3.5.3 to calculate them, respectively.

Whichever method is used, the calculation is simple. Therefore, we can consider that the time overhead used to compute an impact factor of the statement instance is roughly equal to T_0.

Table 3. The first five impact sets of all statement instances in Program 1.

Statement Instance	Direct Impact Successors	Value Impact Set	Path Impact Set	Generalized Value Impact Set	Generalized Path Impact Set	Latent Impact Set
s^1_{13,t_1}	$\varnothing$	s^1_{13,t_1}	$\varnothing$	$\varnothing$	$\varnothing$	$\varnothing$
s^1_{11,t_1}	$\varnothing$	$\varnothing$	s^1_{11,t_1}	s^1_{13,t_1}	$\varnothing$	B^1_{11,t,t_1}
s^1_{9,t_1}	$\varnothing$	$\varnothing$	s^1_{9,t_1}	s^1_{13,t_1}	$\varnothing$	B^1_{9,t,t_1}
s^1_{8,t_1}	$\varnothing$	s^1_{8,t_1}	$\varnothing$	$\varnothing$	$\varnothing$	$\varnothing$
s^3_{5,t_1}	$\varnothing$	$\varnothing$	s^3_{5,t_1}	$\varnothing$	$\varnothing$	B^3_{5,t,t_1}
s^2_{7,t_1}	s^3_{5,t_1}	$\varnothing$	s^3_{5,t_1}	$\varnothing$	$\varnothing$	B^3_{5,t,t_1}
s^2_{6,t_1}	$s^1_{8,t_1}, s^1_{11,t_1}$	s^1_{8,t_1}	s^1_{11,t_1}	s^1_{13,t_1}	$\varnothing$	B^1_{11,t,t_1}
s^2_{5,t_1}	$\varnothing$	$\varnothing$	s^2_{5,t_1}	s^1_{8,t_1}	s^1_{11,t_1}	$\varnothing$
s^1_{7,t_1}	$s^2_{5,t_1}, s^2_{6,t_1}, s^2_{7,t_1}$	s^1_{8,t_1}	$s^2_{5,t_1}, s^3_{5,t_1}, s^1_{11,t_1}$	$s^1_{8,t_1}, s^1_{13,t_1}$	s^1_{11,t_1}	$B^3_{5,t,t_1}, B^1_{11,t,t_1}$
s^1_{6,t_1}	s^2_{6,t_1}	s^1_{8,t_1}	s^1_{11,t_1}	s^1_{13,t_1}	$\varnothing$	B^1_{11,t,t_1}
s^1_{5,t_1}	$\varnothing$	$\varnothing$	s^1_{5,t_1}	s^1_{8,t_1}	s^1_{11,t_1}	$\varnothing$
s^1_{4,t_1}	s^1_{6,t_1}	s^1_{8,t_1}	s^1_{11,t_1}	s^1_{13,t_1}	$\varnothing$	B^1_{11,t,t_1}
s^1_{2,t_1}	$s^1_{5,t_1}, s^1_{6,t_1}, s^1_{7,t_1}$	s^1_{8,t_1}	$s^1_{5,t_1}, s^2_{5,t_1}, s^3_{5,t_1}, s^1_{11,t_1}$	$s^1_{8,t_1}, s^1_{13,t_1}$	s^1_{11,t_1}	$B^3_{5,t,t_1}, B^1_{11,t,t_1}$
s^1_{1,t_1}	$\varnothing$	$\varnothing$	s^1_{1,t_1}	s^1_{8,t_1}	$s^1_{5,t_1}, s^2_{5,t_1}, s^3_{5,t_1}, s^1_{11,t_1}$	B^1_{1,f,t_1}
s^1_{12,t_2}	$\varnothing$,	s^1_{12,t_2}	$\varnothing$	$\varnothing$	$\varnothing$	$\varnothing$
s^1_{11,t_2}	$\varnothing$,	$\varnothing$	s^1_{11,t_2}	s^1_{12,t_2}	$\varnothing$	B^1_{11,f,t_2}
s^1_{9,t_2}	$\varnothing$	$\varnothing$	s^1_{9,t_2}	s^1_{12,t_2}	$\varnothing$	B^1_{9,t,t_2}
s^1_{8,t_2}	$\varnothing$	s^1_{8,t_2}	$\varnothing$	$\varnothing$	$\varnothing$	$\varnothing$
s^1_{5,t_2}	$\varnothing$	$\varnothing$	s^1_{5,t_2}	$\varnothing$	$\varnothing$	B^1_{5,t,t_2}
s^1_{4,t_2}	$s^1_{8,t_2}, s^1_{11,t_2}$	s^1_{8,t_2}	s^1_{11,t_2}	s^1_{12,t_2}	$\varnothing$	B^1_{11,f,t_2}
s^1_{3,t_2}	s^1_{5,t_2}	$\varnothing$	s^1_{5,t_2}	$\varnothing$	$\varnothing$	B^1_{5,t,t_2}
s^1_{1,t_2}	$\varnothing$	$\varnothing$	s^1_{1,t_2}	$\varnothing$	s^1_{5,t_2}	B^1_{5,t,t_2}
s^1_{10,t_3}	$\varnothing$	s^1_{10,t_3}	$\varnothing$	$\varnothing$	$\varnothing$	$\varnothing$
s^1_{9,t_3}	$\varnothing$	$\varnothing$	s^1_{9,t_3}	s^1_{10,t_3}	$\varnothing$	B^1_{9,f,t_3}
s^1_{8,t_3}	$\varnothing$	s^1_{8,t_3}	$\varnothing$	$\varnothing$	$\varnothing$	$\varnothing$
s^3_{5,t_3}	$\varnothing$	$\varnothing$	s^3_{5,t_3}	$\varnothing$	$\varnothing$	B^3_{5,t,t_3}
s^2_{7,t_3}	s^3_{5,t_3}	$\varnothing$	s^3_{5,t_3}	$\varnothing$	$\varnothing$	B^3_{5,t,t_3}
s^2_{6,t_3}	s^1_{8,t_3}	s^1_{8,t_3}	$\varnothing$	$\varnothing$	$\varnothing$	$\varnothing$
s^2_{5,t_3}	$\varnothing$	$\varnothing$	s^2_{5,t_3}	s^1_{8,t_3}	$\varnothing$	$\varnothing$
s^1_{7,t_3}	$s^2_{5,t_3}, s^2_{6,t_3}, s^2_{7,t_3}$	s^1_{8,t_3}	s^2_{5,t_3}, s^3_{5,t_3}	s^1_{8,t_3}	$\varnothing$	B^3_{5,t,t_3}
s^1_{6,t_3}	s^2_{6,t_3}	s^1_{8,t_3}	$\varnothing$	$\varnothing$	$\varnothing$	$\varnothing$
s^1_{5,t_3}	$\varnothing$	$\varnothing$	s^1_{5,t_3}	s^1_{8,t_3}	$\varnothing$	$\varnothing$
s^1_{4,t_3}	s^1_{6,t_3}	s^1_{8,t_3}	$\varnothing$	$\varnothing$	$\varnothing$	$\varnothing$
s^1_{3,t_3}	$s^1_{5,t_3}, s^1_{6,t_3}, s^1_{7,t_3}$	s^1_{8,t_3}	$s^1_{5,t_3}, s^2_{5,t_3}, s^3_{5,t_3}$	s^1_{8,t_3}	$\varnothing$	B^3_{5,t,t_3}
s^1_{1,t_3}	$\varnothing$	$\varnothing$	s^1_{1,t_3}	s^1_{8,t_3}	$s^1_{5,t_3}, s^2_{5,t_3}, s^3_{5,t_3}$	B^1_{1,t,t_3}

Table 4. The first five impact sets of all statements in Program 1.

Statement instance	Value Impact Set	Path Impact Set	Generalized Value Impact Set	Generalized Path Impact Set	Latent Impact Set
s_1	$\varnothing$	$s^1_{1,t_1}, s^1_{1,t_2}, s^1_{1,t_3}$	s^1_{8,t_1}, s^1_{8,t_3}	$s^1_{5,t_1}, s^2_{5,t_1}, s^3_{5,t_1}, s^1_{11,t_1}, s^1_{5,t_2}, s^1_{5,t_3}, s^2_{5,t_3}, s^3_{5,t_3}$	$B^1_{1,f,t_1}, B^1_{5,t,t_2}, B^1_{1,t,t_3}$
s_2	s^1_{8,t_1}	$s^1_{5,t_1}, s^2_{5,t_1}, s^3_{5,t_1}, s^1_{11,t_1}$	$s^1_{8,t_1}, s^1_{13,t_1}$	s^1_{11,t_1}	$B^3_{5,t,t_1}, B^1_{11,t,t_1}$
s_3	s^1_{8,t_3}	$s^1_{5,t_2}, s^1_{5,t_3}, s^2_{5,t_3}, s^3_{5,t_3}$	s^1_{8,t_3}	$\varnothing$	$B^1_{5,f,t_2}, B^3_{5,t,t_3}$
s_4	$s^1_{8,t_1}, s^1_{8,t_2}, s^1_{8,t_3}$	$s^1_{11,t_1}, s^1_{11,t_2}$	$s^1_{13,t_1}, s^1_{12,t_2}$	$\varnothing$	$B^1_{11,t,t_1}, B^1_{11,f,t_2}$
s_5	$\varnothing$	$s^3_{5,t_1}, s^2_{5,t_1}, s^1_{5,t_1}, s^1_{5,t_2}, s^3_{5,t_3}, s^2_{5,t_3}, s^1_{5,t_3}$	s^1_{8,t_1}, s^1_{8,t_3}	s^1_{11,t_1}	$B^3_{5,t,t_1}, B^1_{5,t,t_2}, B^3_{5,t,t_3}$
s_6	s^1_{8,t_1}, s^1_{8,t_3}	s^1_{11,t_1}	s^1_{13,t_1}	$\varnothing$	B^1_{11,t,t_1}
s_7	s^1_{8,t_1}, s^1_{8,t_3}	$s^3_{5,t_1}, s^2_{5,t_1}, s^1_{11,t_1}, s^3_{5,t_3}, s^2_{5,t_3}$	$s^1_{8,t_1}, s^1_{13,t_1}, s^1_{8,t_3}$	s^1_{11,t_1}	$B^3_{5,t,t_1}, B^1_{11,t,t_1}, B^3_{5,t,t_3}$
s_8	$s^1_{8,t_1}, s^1_{8,t_2}, s^1_{8,t_3}$	$\varnothing$	$\varnothing$	$\varnothing$	$\varnothing$
s_9	$\varnothing$	$s^1_{9,t_1}, s^1_{9,t_2}, s^1_{9,t_3}$	$s^1_{13,t_1}, s^1_{12,t_1}, s^1_{10,t_3}$	$\varnothing$	$B^1_{9,t,t_1}, B^1_{9,t,t_2}, B^1_{9,f,t_3}$
s_{10}	s^1_{10,t_3}	$\varnothing$	$\varnothing$	$\varnothing$	$\varnothing$
s_{11}	$\varnothing$	$s^1_{11,t_1}, s^1_{11,t_2}$	$s^1_{13,t_1}, s^1_{12,t_2}$	$\varnothing$	$B^1_{11,t,t_1}, B^1_{11,f,t_2}$
s_{12}	s^1_{12,t_2}	$\varnothing$	$\varnothing$	$\varnothing$	$\varnothing$
s_{13}	s^1_{13,t_1}	$\varnothing$	$\varnothing$	$\varnothing$	$\varnothing$

Table 5. The first five impact factors of all statements in Program 1.

Statement	Value Impact Factor	Path Impact Factor	Generalized Value Impact Factor	Generalized Path Impact Factor	Latent Impact Factor
s_1	0	3	2	8	3
s_2	1	4	2	1	2
s_3	1	4	1	0	2
s_4	3	2	2	0	2
s_5	0	7	2	1	3
s_6	2	1	1	0	1
s_7	2	5	3	1	3
s_8	3	0	0	0	0
s_9	0	3	3	0	3
s_{10}	1	0	0	0	2
s_{11}	0	2	2	0	2
s_{12}	1	0	0	0	0
s_{13}	1	0	0	0	0

Table 6. The execution number of each statement in Program 1.

s_1	s_2	s_3	s_4	s_5	s_6	s_7	s_8	s_9	s_{10}	s_{11}	s_{12}	s_{13}
3	1	2	3	7	4	4	3	3	1	2	1	1

Table 7. Computation for information hidden factor of each statement in Program 1.

Statement	Ratio (fac = 6 class 1)	Ratio (fac = 1 class 2)	Information Hidden Factor
s_1	2/3	1/3	$-\left(\frac{2}{3}\right)\log_2\left(\frac{2}{3}\right) - \left(\frac{1}{3}\right)\log_2\left(\frac{1}{3}\right) = 0.9182$
s_2	1/1		$-\left(\frac{1}{1}\right)\log_2\left(\frac{1}{1}\right) = 0$
s_3	1/2	1/2	$-\left(\frac{1}{2}\right)\log_2\left(\frac{1}{2}\right) - \left(\frac{1}{2}\right)\log_2\left(\frac{1}{2}\right) = 1.0$
s_4	2/3	1/3	$-\left(\frac{2}{3}\right)\log_2\left(\frac{2}{3}\right) - \left(\frac{1}{3}\right)\log_2\left(\frac{1}{3}\right) = 0.9182$
s_5	2/3	1/3	$-\left(\frac{2}{3}\right)\log_2\left(\frac{2}{3}\right) - \left(\frac{1}{3}\right)\log_2\left(\frac{1}{3}\right) = 0.9182$
s_6	2/2		$-\left(\frac{2}{2}\right)\log_2\left(\frac{2}{2}\right) = 0$
s_7	2/2		$-\left(\frac{2}{2}\right)\log_2\left(\frac{2}{2}\right) = 0$
s_8	2/3	1/3	$-\left(\frac{2}{3}\right)\log_2\left(\frac{2}{3}\right) - \left(\frac{1}{3}\right)\log_2\left(\frac{1}{3}\right) = 0.9182$
s_9	2/3	1/3	$-\left(\frac{2}{3}\right)\log_2\left(\frac{2}{3}\right) - \left(\frac{1}{3}\right)\log_2\left(\frac{1}{3}\right) = 0.9182$
s_{10}	1/1		$-\left(\frac{1}{1}\right)\log_2\left(\frac{1}{1}\right) = 0$
s_{11}	1/2	1/2	$-\left(\frac{1}{2}\right)\log_2\left(\frac{1}{2}\right) - \left(\frac{1}{2}\right)\log_2\left(\frac{1}{2}\right) = 1.0$
s_{12}		1/1	$-\left(\frac{1}{1}\right)\log_2\left(\frac{1}{1}\right) = 0$
s_{13}	1/1		$-\left(\frac{1}{1}\right)\log_2\left(\frac{1}{1}\right) = 0$

In the second step, we can conclude that the time overhead used to calculate all factors of all statement instances is roughly equal to five times $\sum_{g=1}^{G}\sum_{k=1}^{K} H_{gk}T_0$, where G is the total number of statements in program under testing, K is the total number of test cases in test suite, and H_{gk} is the number of times statement s_g is executed by the test case t_k. Because statement s_g generates $\sum_{k=1}^{K} H_{gk}$ execution instances, in terms of the program under testing, the total number of executed statement instances by the tests suite is $\sum_{g=1}^{G}\sum_{k=1}^{K} H_{gk}$. Combining with the conclusion in the first step, we get the conclusion: In terms of the program under testing, the time overhead for computing all features of all statement instances is roughly equal to five times $\sum_{g=1}^{G}\sum_{k=1}^{K} H_{gk}$.

In the third step, we can conclude that the time overhead for direct mutation testing is $\sum_{g=1}^{G}\sum_{k=1}^{K} n_g|P_k|T_0$, where we suppose that the statement s_g generates n_g mutants, and the test case t_k executes $|P_k|$ statement instances. In the direct mutation testing, the program under testing generates $\sum_{g=1}^{G} n_g$ mutants, each mutant is tested by the test suite, and the time overhead for the test suite to test each mutant is $\sum_{k=1}^{K} |P_k|T_0$. Therefore, the time overhead for direct mutation testing is $\sum_{g=1}^{G} n_g \times \sum_{k=1}^{K} |P_k|T_0 = \sum_{g=1}^{G}\sum_{k=1}^{K} n_g|P_k|T_0$.

In the fourth step, we compare the time overhead used to calculate all features of all statement instances and the overhead used in the direct mutation testing. The ratio of the two time overheads is $5\sum_{g=1}^{G}\sum_{k=1}^{K} H_{gk}/\sum_{g=1}^{G}\sum_{k=1}^{K} n_g|P_k|$. Because $n_g \gg 5$, $|P_k| \gg H_{gk}$, we can get the final conclusion: Compared with the time required for direct mutation testing, the time overhead used to calculate all statement features can be neglected.

5. Machine Learning Algorithms Comparison and Modelling

Taking the Brier scores as a criterion, we compared the prediction effects of the following five models on statement mutation scores: artificial neural networks (ANN), logical regression (LR), random forests (RF), support vector machines (SVM) and symbolic regression (SR). The experiment result shows the artificial neural network algorithm has the highest prediction precision.

We did not try very complex models because the model should not be too complicated. First, our sample size should not be too large. Our data records need to be extracted in real time, so that the excessively large sample size will cause the user to wait a long time. In the case of a small sample size, over-complexing models can cause over-fitting. Secondly, according to the introduction of the Section 3, we can know that the relationship between the dependent variable and each independent variable is monotonic, so we estimate that the available model should not be very complicated.

5.1. Experimental Subjects

In this paper, there are two programs under testing: schedule.c and tcas.c. We explain our experiment with schedule.c as the main part and tcas.c as the auxiliary part. The program schedule.c realizes a CPU process management, and the program tcas.c realizes an aircraft early warning system. A more specific introduction is as follows.

The program schedule.c [16] realizes a priority scheduling algorithm. A computer has only one CPU, but sometimes multiple programs simultaneously request to be executed. For solving this problem, the priority scheduling algorithm assigns each program a priority. When a program needs to use CPU, it is first stored in a queue so that the program with a higher priority gets a CPU, whereas the program with a lower priority can wait. The schedule.c consists of 73 lines of C code including one branch statement, two single-loop statements and two double-loop statements. The test cases are included in its usage instructions. We take these test cases as a test suite of schedule.c.

The program tcas.c [17] is used to avoid collision of aircraft, which consists of 135 lines of C code with 40 program branch statements and 10 compound predicates. The tcas.c is able to monitor the traffic situation around a plane and offer information on the altitude of other aircraft. It can also generate collision warnings that another aircraft is in close vicinity by calculating the vertical and horizontal distances between the two aircrafts. The Software artifact Infrastructure Repository (SIR) also supplies some types of test case suites for tcas.c. From the SIR, we randomly selected a branch coverage test suite $suite122$ as the test suite used in our experiment.

5.2. The Construction Method of Data Set

To compare the prediction accuracy of the five machine learning models, we did two experiments with schedule.c and tcas.c, respectively. No matter the experiment, the data set is created in the same way. In each experiment, the data set contains 200 data records. Each data record r_p is established with one corresponding mutant sample m_p and contains seven independent variables and one dependent variable. If m_p is generated by modifying the statement s_q, then the seven independent variables of r_p are the seven features of the statement s_q, and the dependent variable of r_p is the identification result of the mutant m_p.

We take an example to explain the construction process of a data record. We might as well assume that a mutant sample m_p is generated by modifying statement s_2 and identified by the test suite. Now, we use m_p to construct one data record r_p. Because m_p is generated by modifying s_2, the values of seven independence variables in r_p are the seven features of statement s_2, i.e., $(1,4,2,1,2,1,0)$ as shown in Tables 5–7. Because m_p is identified by the test suite, the value of dependence variable in r_p is 1. Therefore, the data record r_p is $(1,4,2,1,2,1,0,1)$.

5.3. Performance Metrics

A model may be considered good when it is evaluated using a metric, but, at the same time, the model may be considered bad when assessed against other metrics. For this reason, we will compare a few different common evaluation metrics and decide which of them is more suitable to our statement mutation score prediction.

5.3.1. Area under Curve

The two coordinates of the receiver operating characteristic (ROC) curve represent sensitivity and specificity, respectively. Through these two indicators, the ROC curve displays the two types of errors for all possible thresholds. The area AUC under the ROC curve is the quantitative indicator commonly used to evaluate a binary classification algorithm [18].

5.3.2. Logarithmic Loss

Logarithmic Loss works by penalising the false classifications [18]. It works well for both binary classification and multi-class classification generally. For a binary classification, the logarithmic function

$$-\frac{1}{n}\sum_{i=1}^{n} I(y_i = 1) \log \left[\hat{p}(Y = 1|x_i)\right] + I(y_i = 0) \log \left[1 - \hat{p}(Y = 1|x_i)\right]$$

is often used as a classifier's loss function. Logarithmic Loss closer to 0 indicates higher accuracy for the classifier.

5.3.3. Brier Score

The basic idea of Brier score is to compute the mean squared error (MSE) between the predicted probability scores and the true class indicator [19], where the positive class is coded as $y_i = 1$, and negative class $y_i = 0$. The most common formulation of the Brier score is shown as follows:

$$BS = \frac{1}{n}\sum_{i=1}^{n}[y_i - \hat{p}(Y = y_i|x_i)]^2.$$

The Brier score is a loss function, which means the lower its value, the better the machine learning model.

5.3.4. Metric Comparison

In the cross-validation process, we choose the Brill score as the model evaluation criterion. Our purpose is only to tell our users how likely the software bug in a statement will be detected by a test suite. Therefore, AUC is not suitable for us because it is also not directly related to the predicted probability. Because the logarithmic loss function may lead to an infinite penalty, it is also not used by us. The Brier score is the good score function because it is related to the predicted probability and is bounded. For the above reasons, we take the Brier score as an evaluation criterion in the cross-validation.

5.4. Model Comparing and Tuning

Under the condition of the same partitioning of the data set, we take the Brier score as a standard to evaluate the model. In our experiment, we tune hyperparameters and compare the prediction accuracies of five machine learning models. We use the same partitioning of the data set and the repeated 5-fold cross-validation to evaluate the prediction accuracy of the models because of the two following reasons.

(1) We tune some hyperparameters to find the optimal model settings with the help of the repeated 5-fold cross-validation method. During the 5-fold cross-validation, the samples are randomly partitioned into five equally sized folds. Models are then fitted by repeatedly leaving out one of the folds. In our each experiments, our data set contains 200 data records, so that the training and validation sets contain 160 and 40 data records, respectively. However the result from cross-validation is more or less uncertain generally. Therefore, in our experiment, five repeats of 5-fold cross-validation are used to effectively reduce this uncertainty and increase the precision of the estimates. Because each

5-fold cross-validation supplies a Brier score, five repeats of 5-fold cross-validation supply 5 Brier scores. Under each candidate combination of hyperparameters, we use the average of the five Brier scores to represent the prediction effects of the corresponding model.

(2) Because the performance metric is sensitive to the data splits, we thus compare the machine learning models based on the same partitioning of the data. Otherwise, the difference in performance will come from two different sources: the differences among the data splits and the differences among the models themselves. If one model is better than the other, we don't know if all performance differences are caused by model differences.

The compared models include the logistic regression, random forest, neural network, support vector machine and symbolic regression. We use their average Brier scores to assess their prediction effects.

5.4.1. Logistic Regression

(1) Introduction to Logistic Regression

Conventional logistic regression [20,21] can predict the occurrence probability of a specific outcome. The conditional probability of a positive outcome could be expressed with the formula below:

$$p(x_i) = p(Y = 1|x_i) = \frac{1}{1 + e^{\beta_0 + \beta_1 x_1 + \beta_2 x_2 + \cdots + \beta_d x_d}},$$

where β_i is the coefficient for the ith feature, and d is the total number of features. $\beta_1, \beta_2, \cdots, \beta_d$ can be solved by the elastic net approach [22–24] as follows:

$$max_{\beta_0,\beta} \frac{1}{n} \sum_{i=1}^{n} \{I(y_i = 1) log p(x_i) + I(y_i = 0) log(1 - p(x_i))\} - \lambda \left[(1 - \alpha) \frac{1}{2} \|\beta\|_2^2 + \alpha \|\beta\|_1 \right], \quad (14)$$

where

$$\|\beta\|_2^2 = \beta_1^2 + \beta_2^2 + \cdots + \beta_d^2 \quad and \quad \|\beta\|_1 = |\beta_1| + |\beta_2| + \cdots + |\beta_d|.$$

(2) Logistic regression tuning

Glmnet [25] is an R language software package that can fit linear, logistic and multinomial, Poisson, and Cox regression models by maximizing the penalized likelihood. In order to predict the mutant score of each program statement in schedule.c, we use the ridge penalty algorithm in a glmnet software package to fit the logistic regression mode. Hence, during tuning hyper parameters, the penalized parameter α in the formula (14) is set to 0, and the penalized parameter λ is set to 10^i where i takes each integer from -7 to 7 in turn. In the cross-validation process, we use the Brier score as the model evaluation criterion. Under each penalized parameter, the five repeats of 5-fold cross-validation generate five Brier scores. We calculate the average of the five Brier scores under each candidate penalized parameter λ, so that we can use the average Brier score to represent the prediction effect of the model under the each candidate penalized parameter.

Figure 4 and Table 8 show the average Brier Score under each candidate value of the penalized parameter λ. In Figure 4, the profile shows a decrease in the average Brier score until the penalized value λ is 10^{-2}. Therefore, the numerically optimally value of the penalized parameter is 10^{-2}.

Table 8. Average Brier scores for the logistic regression model.

λ	10^{-7}	10^{-6}	10^{-5}	10^{-4}	10^{-3}
Mean	0.1095	0.1062	0.1075	0.1016	0.0955
λ	10^{-2}	10^{-1}	1	10^{1}	10^{2}
Mean	0.0950	0.1090	0.1321	0.1374	0.1380
λ	10^{3}	10^{4}	10^{5}	10^{6}	10^{7}
Mean	0.1381	0.1381	0.1381	0.1381	0.1381

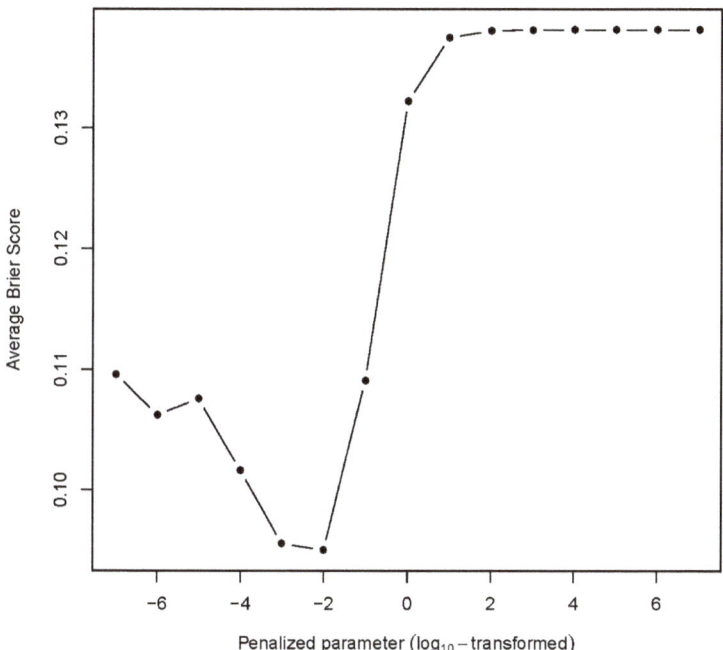

Figure 4. The performance profile of the logistic regression for predicting the statement mutation scores.

5.4.2. Random Forests

(1) Introduction to Random Forest

The random forest model [26,27] can work for regression tasks and classification tasks generally. It is a tree-based model consisting of multiple decision trees. Each decision tree is created on an independent and random sample taken from the training data set.

The decision tree algorithm [18,28] is a top-down "greedy" approach that partitions the dataset into smaller subsets. This algorithm has a tree-like structure that predicts the value of a target variable based on several input variables. At each decision node, the features are split into two subsets and this process is repeated until the number of data in the splits falls below some threshold. According to the target variable's type, decision trees can be divided into regression trees and classification trees. The purpose of classification tree is to classify, and its target variable takes discrete values. The purpose of regression trees is to build a regression model, its target variable takes continuous values.

For regression, the regression tree algorithm begins with the entire data set S and searches every distinct value of every independent variable to find the appropriate independent variable and split its value that partitions the data into two subsets (S_1 and S_2) such that the overall sums of squares error

$$SSE = \sum_{i \in S_1} (y_i - \bar{y}_1)^2 + \sum_{i \in S_2} (y_i - \bar{y}_2)^2 \tag{15}$$

are minimized, where $\bar{y}_1$ and $\bar{y}_2$ are the averages of the outcomes within subsets S_1 and S_2, respectively. Then, within each of subsets S_1 and S_2, this method searches again for the independent variable and splits its value that best reduces SSE. Because of the recursive splitting nature of regression trees, this method is also known as recursive partitioning.

For classification, the aim of classification trees is to partition the data into smaller, more homogeneous groups. Homogeneity in this context means that the nodes of the split are more pure. This purity is usually quantified by the entropy or Gini index. For the two-class problem, the Gini index for a given node is defined as

$$p_1(1 - p_1) + p_2(1 - p_2), \tag{16}$$

where p_1 and p_2 are the probabilities of Class 1 and Class 2, respectively.

In order to make a prediction for a given observation, the regression tree first analyzes which class this observation belongs to, and then takes the mean of the training data in the class as the prediction of this observation. When random forest algorithms are used, the result of regression question can be obtained by averaging predictions across all regression trees, and the result of the classification question can be obtained by a majority vote across all classification trees, respectively. The generalization error of a random forest depends on the errors of individual trees and the correlation between the trees.

(2) Random forest tuning

The randomForest package [29] implements the random forest algorithm in the R environment. We use this software to predict statement mutation scores generated when the test suite executes on the program schedule.c. Because the statement mutation score can be considered as the probability of positive class in binary classification, we denote the positive class and negative class as 1 and 0, respectively, and let Random Forests run under regression mode to predict the probability of a positive class [30]. In order to obtain a good prediction model, the different hyper parameter combinations are tried. The most important hyper parameter is mtry, which is the number of independent variables randomly selected at each split. In our experiment, we tried multiple candidate values of mtry (from 1 to 7). The other important tuning parameter is ntree, which is the number of bootstrap samples in the random forest algorithm. In theory, the performance of a random forest model should be a monotonic function of the number of trees (ntree). However, when ntree is greater than a certain number, the performance of a random forest model can only improve slowly. In our experiment, ntree is set to 1000. Under each candidate value of the parameter mtry, we calculate the average of the five Brier scores generated from the five repeats of 5-fold cross-validation. Furthermore, we use these averages to express the prediction effects of the random forest model under different candidate values of hyperparameter mtry. Figure 5 and Table 9 show the average Brier score under each candidate value of the hyperparameter mtry. As shown in Figure 5, the average Brier scores show a U shape, whose minimum value occurs in $mtry = 3$. Therefore, 3 is the optimal value of mtry.

Table 9. Average Brier scores for the random forest model.

mtry	1	2	3	4	5	6	7
Mean	0.1166	0.0923	0.0888	0.0897	0.0914	0.0928	0.0925

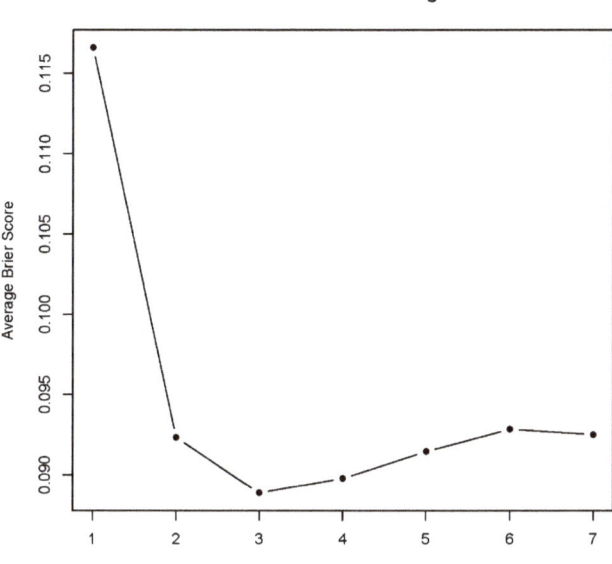

Figure 5. The performance profile of the random forest for predicting the statement mutation scores.

5.4.3. Artificial Neural Networks

(1) Introduction to neural network

Neural networks [18,31] can be used not only for regression but also for classification. The outcome of a neural network is modeled by an intermediary set of unobserved variables called hidden units. The simplest neural network architecture is the single hidden layer feed-forward network.

The working process of the single hidden layer feed-forward neural network is as follows. During the entire work of the neural network, all input neurons representing the original independent variables $x_1, x_2, \cdots, x_s$ are first activated through the sensors perceiving the environment. Next, inside each hidden unit h_k, all original independent variables are linearly combined to generate

$$u_k(x) = \beta_{0k} + \sum_{j=1}^{s} \beta_{jk} x_j, \tag{17}$$

where $k = 1, 2, \cdots, r$ and r is the number of the hidden units. Then, by a nonlinear function g_k, $u_k(x)$ is typically transformed into the output of hidden unit h_k as follows:

$$g_k(x) = \frac{1}{1 + e^{-u_k(x)}}. \tag{18}$$

(i) When treating the neural network as a regression model, all $g_k(x)$ are linearly combined to form the output of neural network:

$$f(x) = \gamma_0 + \sum_{k=1}^{r} \gamma_k g_k(x). \tag{19}$$

All of the parameters β and γ can be solved by minimizing the the penalized sum of the squared residuals:

$$\sum_{i=1}^{n}(y_i - f(x_i))^2 + \lambda \sum_{k=1}^{r}\sum_{j=0}^{s}\beta_{jk}^2 + \lambda \sum_{k=0}^{r}\gamma_k^2, \quad (20)$$

where $f(x_i)$ and y_i are the predicted result and the actual result related to the ith observed data, respectively.

(ii) Neural networks can also be used for classification. Unlike neural networks for regression, an additional nonlinear transformation is used on the linear combination of the outputs of hidden units.

When the neural network is used for binary classification, it uses

$$f^*(x) = \frac{1}{1 + e^{-f(x)}} = \frac{1}{1 + e^{-(\gamma_0 + \sum_{k=1}^{r}\gamma_k g_k(x))}} \quad (21)$$

to predict the class probability. The estimation of the parameters γ and β can be solved by minimizing the penalized cross-entropy

$$-\sum_{i=1}^{n} y_i \log f(x_i) + (1 - y_i)\log(1 - f(x_i)) + \lambda \sum_{k=1}^{r}\sum_{j=0}^{s}\beta_{jk}^2 + \lambda \sum_{k=0}^{r}\gamma_k^2, \quad (22)$$

where y_i is the 0/1 indicator for the positive class. The neural network algorithm can also be used for multi-class classification. In this situation, the softmax transform outputs the probability that the sample x belongs to the lth class. Except the single hidden layer feed-forward network, there are many other types of models. For example, the famous deep learning approaches consist of multiple hidden layers.

(2) Neural network tuning

As we said before, the our model must not be too complicated, so we select R package nnet [32] to predict the statement mutation scores of the test suite on schedule.c. The software package nnet implements a feed-forward neural network with a single hidden layer. The λ and r in formula (22) represent the weight decay and the number of units in the hidden layer, respectively. They are denoted as *decay* and *size* in nnet package, respectively. Therefore, *decay* is the regularization parameter to avoid over-fitting.

In our experiment, *size* is set in turn to each integer value between one and then. At the same time, the *decay* was set to 10^i where i takes each integer value from -4 to 5 in turn.

Figure 6 and Table 10 show the average Brier scores under the each candidate combinations of *size* and *decay*. From them, we can know that the optimal combination of the weight decay and hidden unit number is $decay = 10^{-2}$ and $size = 8$ because, at this time, the minimum average Brier score appears.

Table 10. Average Brier scores for neural network models.

Size \ Decay	10^{-4}	10^{-3}	10^{-2}	10^{-1}	1	10^1	10^2	10^3	10^4	10^5
1	0.1180	0.0984	0.0928	0.0968	0.1111	0.1397	0.1956	0.2419	0.2491	0.2499
2	0.0915	0.0869	0.0883	0.0876	0.1104	0.1379	0.1889	0.2404	0.2489	0.2498
3	0.0953	0.0890	0.0865	0.0867	0.1099	0.1371	0.1835	0.2390	0.2488	0.2498
4	0.0897	0.0889	0.0863	0.0881	0.1094	0.1367	0.1791	0.2375	0.2486	0.2498
5	0.0874	0.0890	0.0865	0.0880	0.1093	0.1366	0.1753	0.2361	0.2484	0.2498
6	0.0881	0.0869	0.0865	0.0881	0.1093	0.1365	0.172	0.2348	0.2483	0.2498
7	0.0896	0.0887	0.0862	0.0878	0.1093	0.1364	0.1694	0.2334	0.2481	0.2498
8	0.0884	0.0878	0.0856	0.0882	0.1093	0.1364	0.1670	0.2321	0.2479	0.2497
9	0.0884	0.0871	0.0869	0.0881	0.1093	0.1364	0.1649	0.2308	0.2478	0.2497
10	0.0870	0.0880	0.0867	0.0878	0.1093	0.1365	0.1631	0.2295	0.2476	0.2497

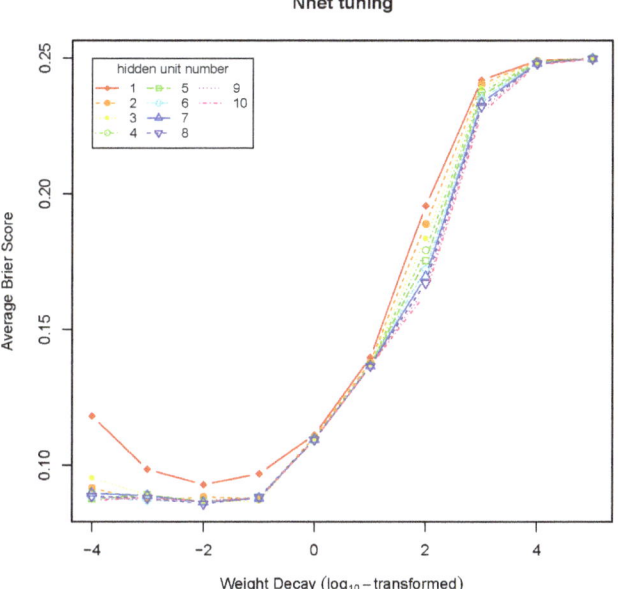

Figure 6. The performance profile of the neural network for predicting the statement mutation scores.

5.4.4. Support Vector Machines

(1) Introduction to support vector machine

Given a set of n training instances $x_1, x_2, \cdots, x_n$, the goal of support vector machine is to find a hyperplane that separates the positive and the negative training instances with the maximum margin and minimum misclassification error. The training of support vector machine is equivalent to solving the following optimization problem:

$$\min_{w,b,\zeta_i} \frac{1}{2}\|w\|^2 + C\sum_{i=1}^{n} \zeta_i$$
$$\text{subject to} \quad y_i(w^T x_i + b) \geq 1 - \zeta_i$$
$$\zeta_i \geq 0, \quad i = 1, 2, \cdots, n,$$

where w is the normal vector of the maximum-margin hyperplane $w^T x + b = 0$, C is the regularization parameter, ζ_i indicates a non-negative slack variable to tolerant some training data falling in the wrong side of the hyperplane, and b is a bias. The parameter C specifies the cost of a violation to the margin. When C is small, the margins will be wide and many support vectors will be on the margin or will violate the margin. When C is large, the margins will be narrow and there will be few support vectors on the margin or violating the margin.

The maximum-margin hyperplane can be obtained by solving the above problem. Given new data x, $f(x) = w^T x + b$ represents the signed distance between x and the hyperplane. We can classify the new data x based on the sign of $f(x)$.

If the original problem is stated in a finite-dimensional space, it often happens that the sets to discriminate are not linearly separable. For solving this problem, a support vector machine maps the original finite-dimensional space into a higher-dimensional space, making the separation easier.

Let $\phi(x)$ denote the vector after x mapping. In this higher-dimensional space, the optimization problem can be rewritten into

$$\min_{w,b,\zeta_i} \frac{1}{2}\|w\|^2 + C\sum_{i=1}^{n} \zeta_i \tag{23}$$

$$\text{subject to} \quad y_i(w^T\phi(x_i) + b) \geq 1 - \zeta_i \tag{24}$$

$$\zeta_i \geq 0, \quad i = 1, 2, \cdots, n \tag{25}$$

or expressed in the dual form

$$\min_{\alpha} \quad -\sum_{i=1}^{n} \alpha_i + \frac{1}{2}\sum_{i=1}^{n}\sum_{j=1}^{n} \alpha_i \alpha_j y_i y_j k(x_i, x_j) \tag{26}$$

$$\text{subject to} \quad \sum_{i=1}^{n} \alpha_i y_i = 0 \tag{27}$$

$$0 \leq \alpha_i \leq C, i = 1, 2, \cdots, n, \tag{28}$$

where $k(x_i, x_j) = \phi(x_i)^T \phi(x_j)$ defines the kernel function greatly reducing the computational cost.

By solving the above optimization problem, the optimal α_i^* and b^* can be obtained. Therefore, the maximum-margin hyperplane in the higher-dimensional space is

$$w^{*T}\phi(x) + b = \sum_{i=1}^{n} \alpha_i^* y_i \phi(x_i)^T \phi(x) + b^* = \sum_{i=1}^{n} \alpha_i^* y_i k(x_i, x) + b^*. \tag{29}$$

The kernel trick allows the support vector machine model to produce extremely flexible decision boundaries. The most common kernel functions are listed in Table 11:

Table 11. Kernel functions.

Name	Expression	Parameter
linear kernel	$k(x_i, x_j) = x_i^T x_j$	
polynomial kernel	$k(x_i, x_j) = (\gamma x_i^T x_j + \theta)^d$	γ, θ, d
radial kernel	$exp(-\sigma\|x_i - x_j\|^2)$	σ

The original SVM can be used for classification and regression without probability information. To solve this problem, Platt [33] proposed to use a logistic function to convert the decision value from a binary support vector machine to a probability. Formally, the probability of data x_i being a positive instance is defined as follows:

$$P(y_i = 1|x_i) = \frac{1}{1 + exp(Af(x_i) + B)},$$

where $f(x) = w^T\phi(x) + b$ is the maximum-margin hyperplane. The parameters A and B are derived by minimizing the negative log-likelihood of the training data:

$$-\sum_{i=1}^{n}[t_i log(P(y_i = 1|x_i)) + (1 - t_i)log(1 - P(y_i = 1|x_i))]$$

where

$$t_i = \begin{cases} \frac{n_+ + 1}{n_+ + 2} & \text{if } y_i = 1, \\ \frac{1}{n_- + 2} & \text{if } y_i = -1. \end{cases}$$

n_+ denotes the number of positive training instances (i.e., $y_i = 1$), and n_- denotes the number of negative training instances (i.e., $y_i = -1$). Newton's method with backtracking is a commonly used approach to solve the above optimization problem [34] and is implemented in LibSVM. Besides the binary classification, the support vector machine can also compute the class probabilities for the multi-class problem using one-against-one (i.e., pairwise) approach [35].

(2) Support vector machine tuning

Support vector machine algorithms are provided in the software package kernlab [36] written in the R language. We built the support vector machine based on the radial basis kernel function provided by this package. A radial basis kernel function maps the independent variables to an infinite-dimensional space. The regularization parameter C in formula (23) is called cost parameter in kernlab. A smaller C results in a smoother decision surface and a larger C results in a flexible model that strives to classify all training data correctly. The radial basis kernel function in kernlab package is shown in Table 11, where the parameter σ represents the inverse kernel width. A larger σ means a narrower radial basis kernel function.

When we use kernlab to predict the statement mutation scores of schedule.c, we hope to get the Brier score as small as possible by tuning C and σ. For this purpose, we first set the parameter σ to the median of $\|x - x'\|^2$ [18,37,38]. Next, let the parameter C take respectively as $2^{-5}, 2^{-3}, 2^{-1}, 2^1, 2^3, 2^5, 2^7, 2^9, 2^{11}, 2^{13}$ and 2^{15}. Then, under each candidate value of C, we use the five repeats of 5-fold cross-validation to calculate the average Brier scores.

Figure 7 and Table 12 show the average Brier score generated by five repeats of 5-fold cross-validation at each candidate value of C. As shown in Figure 7, although there was a relatively large fluctuation, the average Brier score shows a general trend of first decreasing and then rising. From this figure, we can know that $C = 2^{-1}$ is the optimally value of the regularization parameter. At this time, the average Brier score reaches a minimum 0.0933.

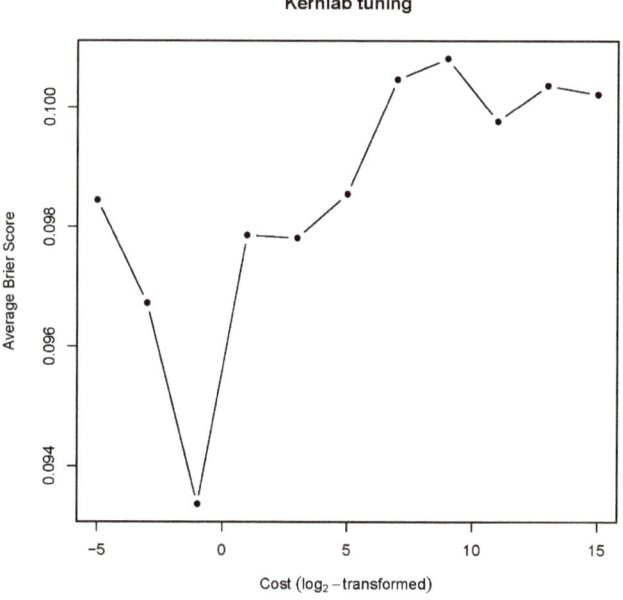

Figure 7. The performance profile of the support vector for predicting the statement mutation scores.

Table 12. Resampled Brier score for the support vector machine model.

C	2^{-5}	2^{-3}	2^{-1}	2^1	2^3	2^5	2^7	2^9	2^{11}	2^{13}	2^{15}
Mean	0.0984	0.0967	0.0933	0.0978	0.0978	0.0985	0.1004	0.1008	0.0997	0.1003	0.1002

5.4.5. Symbolic Regression

(1) Introduction to symbolic regression

Symbolic regression can also be called function modeling. Based on the given data, it can automatically find the functional relationship, such as $2x^3 + 5$, $\cos(x) + 1/e^y$, etc., between independent variables and dependent variables.

Throughout the modeling process, a function model $f(x)$ is always coded as a symbolic expression tree. The input of symbolic regression is a data set, and the genetic programming method is often used to determine $f(x)$. The genetic programming constantly changes an old function model into a new better fitted one by selecting the function with the better fitness value. A possible and frequently used fitness function is the average squared difference between the values predicated by $f(x)$ and the actually observed values y as follows:

$$MSE(f(x), y) = \frac{1}{n}\sum_{i=1}^{n}(f(x_i) - y_i)^2.$$

Mutation operations and crossover operations are the two important ways to change function model $f(x)$. A mutation operation directly changes a symbolic expression sub tree, and a crossover operation cuts a symbolic expression sub tree and replace it with a sub tree in another symbolic expression tree.

(2) Symbolic regression tunning

The symbolic regression tool rgp [39] is an implementation of genetic programming methods in the R environment. We use rgp to predict the statement mutation scores of schedule.c. In our symbolic regression experiment, the most basic mathematical operators are set to the operators +, −, *, /, sin. An important tuning parameter in rgp is *populationSize*, which means the number of individuals included in a population, and is set to 100 in our experiment.

Another important tuning parameter is the number of evolution generations. Too few evolutionary generations produce an under-fitting, whereas too many evolutionary generations produce an over-fitting. We did a grid search to determine the optimal number of the generations, which minimizes the average Brier score. Because we need to complete the model fitting in a relatively short time, the number of evolution generations cannot be set too large. In our experiment, the candidate number of evolution generations is set to 3, 6, 9, 12, 15, 18, respectively. The five repeats of five-fold cross-validation are used to calculate the evolution effects (i.e., the average Brier scores) under each candidate number of evolution generations.

As shown in Table 13 and Figure 8, the average Brier scores oscillated down. In the 12th generation evolution, the smallest average Brier score 0.1504 appeared.

Table 13. Average Brier scores for the symbolic regression.

Generation	3	6	9	12	15	18
Mean	0.1640	0.1546	0.1548	0.1504	0.1527	0.1515

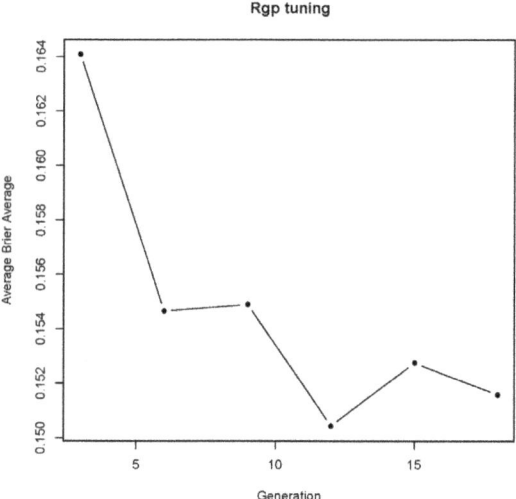

Figure 8. Rgp tuning.

5.4.6. Comparing Models

Once the hyper parameters in the above five models have been determined for the above five models, we face the question: how do we choose among multiple models? The logistic regression model is used to set the baseline performance because its mathematical expression is simple and operation speed is fast. If other predictive models do not surpass it, the logistic regression model is used in future actual forecasting.

The boxplot in Figure 9 shows, under the condition that the Brier score is the standard, the neural network does the best job about predicting the statement mutation scores. The second best is the random forest model, which is a little better than the support vector machine model. The logistic model is second to last and greatly exceeded the symbolic regression.

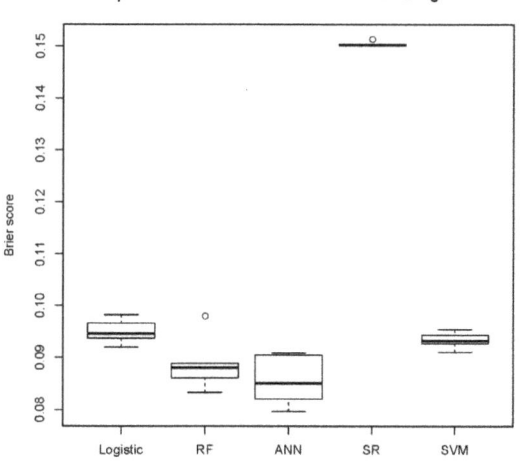

Figure 9. Comparison of the Brier scores of the five machine learning models for schedule.c.

5.4.7. Testing Predictions in Practice

According to Figure 9 and Tables 8–10, 12 and 13, we know that, in the process of repeated cross-validation, the average Brier scores of the logistic regression, random forest, neural network, support vector machine and symbolic regression are 0.0950, 0.0888, 0.0856, 0.0933 and 0.1504, respectively. Therefore, the neural network is the best model because its average Brier score is lower than other models. To further demonstrate the predictive effect of the neural network model on the schedule.c, we did the two following things. Firstly, we apply the neural network model, whose hyper-parameters have been tuned according to the method in Section 5.4.3, to predict the statement mutation scores of schedule.c. Under the condition that the schedule.c is used as the experiment subject, we calculate the mean absolute error between all statement mutation scores obtained by the neural network prediction and all real statement mutation scores. The experiment result shows the mean absolute error reaches 0.1205. Secondly, we randomly select 34 statements in schedule.c, and their two kinds of statement mutation scores are shown in Figure 10. In this figure, the horizontal coordinate represents the real statement mutation score, and the vertical coordinate represents the statement mutation score predicted by the neural network model. Each circle represents a statement, and   From this figure, we can see that :   ashed lines.

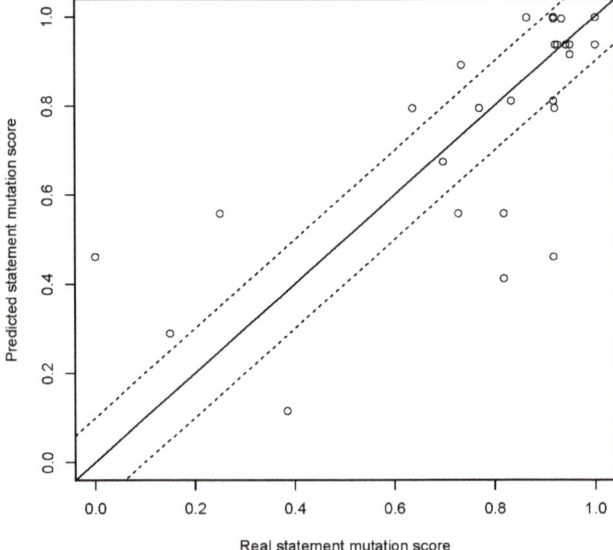

Figure 10. Comparing the real statement mutation scores and the predicted statement mutation scores in schedule.c by the artificial neural network.

5.5. *Further Confirmation of the Optimization Model*

To further confirm that the prediction effect of the neural network is the best, we compared the five machine learning models again under the condition that the program tcas.c is used as the experimental subject. In the process of repeated cross-validation, the average Brier score of the neural network model reaches 0.1164. The average Brier scores of the logistic regression, the support vector machine, the random forest and the symbolic regression are 0.1233, 0.1249, 0.1289 and 0.1373, respectively. Therefore, the neural network is once again considered the best model because its average Brier score is lower than other models. To further demonstrate the predictive effect of the neural network model on the tcas.c, we apply the neural network model, whose hyper-parameters have been tuned according to the method in Section 5.4.3, to predict the statement mutation scores of tcas.c. Under the condition that the tcas.c is used as the experiment subject, the mean absolute error between real statement mutation scores and the statement mutation scores predicted by the tuned neural network reaches 0.1198. In order to illustrate the prediction results of the neural network more vividly, we randomly selected 31 statements in the program tcas.c. Their real statement mutation scores and the corresponding predicted mutation scores are show:

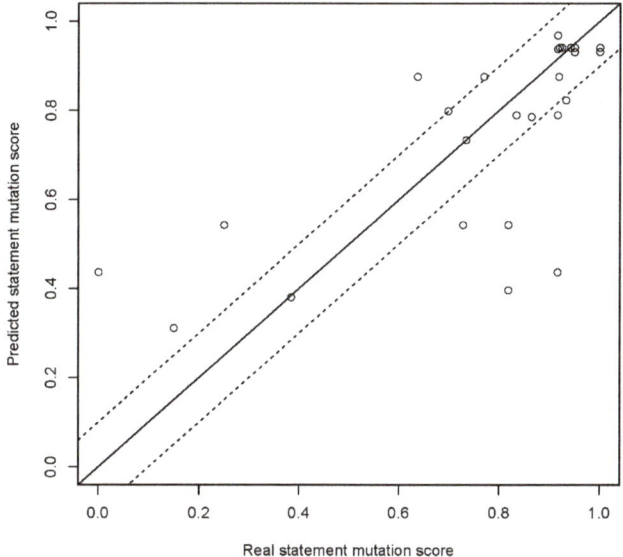

Figure 11. Comparing the real statement mutation scores and the predicted statement mutation scores in tcas.c by the artificial neural network.

Through the above analysis, we can see that, whether the experiment subject is schedule.c or tcas.c, the average Brier scores of the neural network are both the minimums. Thus, we recommend the single hidden layer feedforward neural network as the best model. In the two experiments, the mean absolute error between the statement mutation scores predicted by the neural network model and the real statement mutation scores both approximately reach 0.12.

6. Structure of the Automated Prediction Tool

The work process of our automatic analysis tool consists of the five parts, as shown in Figure 12: extracting the features of the statements in the program under testing, generating mutants, executing test suite on the each mutants, establishing the neural network model, and predicting the statement mutation scores.

In the first part, we extract the features of statements in the program under testing. First, we execute each test case and construct its execution impact graph with the open source software giri [40]. Giri was originally a dynamic program slice tool and is currently modified by us. Next, we traverse the statement instances in reverse order of the execution history of the test cases. Whenever we visit a statement instance, we compute its features. After calculating the features of each statement instance, we calculate the features of each program statement according to the corresponding the statement instances.

In the second part, we generate mutants. We first build a mutation operator set. In our experiments, the mutation operator set consists of the 22 mutation operators, which exist in the open source mutant generate tool ProteumIM2.0 [12]. These operators include u-Cccr, u-OEAA, u-OEBA, u-OESA, u-CRCR, u-Ccsr, u-OAAN, u-OABN, u-OALN, u-OARN, u-OASN, u-OCNG, u-OLAN, u-OLBN, u-OLLN, u-OLNG, u-OLRG, u-OLSN, u-ORBN, u-ORLN, u-ORRN and u-ORSN. Next, we use these mutation operators to randomly construct 200 mutants, each of which is the program with a software bug.

In the third part, we execute the test suite on each mutant and record the corresponding identification result.

In the fourth part, we take the features of the mutant as independent variables and the identification result of the mutant as dependent variables to construct the prediction model with the neural network.

In the fifth part, we predict the mutation scores of each program statement with the constructed model.

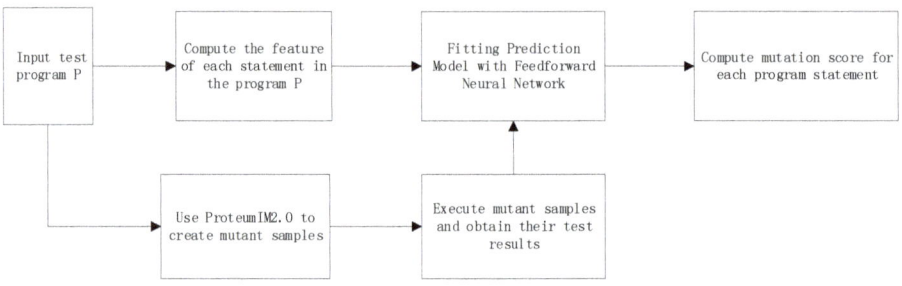

Figure 12. The structure of Automated Analysis Tool.

7. Conclusions

In this paper, we predicted statement mutation scores while using a single hidden layer feedforward neural network and seven statement features. As analyzed in Section 5, each experimental result shows that the neural network is the best prediction model from the standpoint of the mean absolute error. The experimental results on two c programs demonstrate that our method can directly predict statement mutant scores approximately. The experiment results also show that the seven statement features that represent the dynamic program execution and testing process can basically reflect the impact of statements on program output.

However, two shortcomings need to be improved. Firstly, a part of statement features weakly related to program outputs still need to be discovered. If the real mutation score of a statement is low, then this statement usually only has some statement features weakly related to program outputs. In this case, the prediction effect of my model is not good because we only found a part of weakly relevant statement features, and the other part of the weakly correlated sentence features still need to be discovered.

Secondly, in this paper, we assume the controlling expression has no side effect. However, in a few cases, a controlling expression has a side effect. In this case, the execution instance of the controlling expression may have some impact successors. For example, if there is a controlling expression if(x > y++) in the original program, then the execution result of y++ will be changed when it is executed a test case, so that it impacts subsequent statement instances containing y variables. In this situation, the methods mentioned in Sections 3.1.4, 3.2.4, 3.3.2, 3.4.2 and 3.5.3 are no longer applicable, and the corresponding algorithm needs to be redesigned.

In the future, we also plan to predict the statement mutation scores with the prediction model established by other programs. In this case, the users can train a prediction model with the data records from other programs beforehand. Using this pre-trained model, users can directly predict the statement mutation scores of the current program.

Author Contributions: Conceptualization, L.T.; methodology, L.T.; Supervision, Y.W. and Y.G.; writing—original draft preparation, L.T.

Funding: This work was supported by the National Natural Science Foundation of China (No. U1736110), the National Natural Science Foundation of China (No. 61702044), and the Fundamental Research Funds for the Central Universities (No. 2017RC27).

Conflicts of Interest: The authors declare no conflicts of interest.

Abbreviations

The following abbreviations are used in this manuscript:

V_g	value impact set of the statement s_g
$x_{vi}(s_g)$	value impact factor of the statement s_g
V_{g,t_k}^h	value impact set of the statement instance s_{g,t_k}^h
V_{r,z,t_k}^l	value impact set of the branch instance B_{r,z,t_k}^l
P_g	path impact set of the statement s_g
$x_{pi}(s_g)$	path impact factor of the statement s_g
P_{g,t_k}^h	path impact set of the statement instance s_{g,t_k}^h
P_{r,z,t_k}^l	path impact set of the branch instance B_{r,z,t_k}^l
$\mathcal{V}_g$	generalized value impact set of the statement s_g
$x_{gvi}(s_g)$	generalized value impact factor of the statement s_g
$\mathcal{V}_{g,t_k}^h$	generalized value impact set of the statement instance s_{g,t_k}^h
$\mathcal{P}_g$	generalized path impact set of the statement s_g
$x_{gpi}(s_g)$	generalized path impact factor of the statement s_g
$\mathcal{P}_{g,t_k}^h$	path impact set of the generalized statement instance s_{g,t_k}^h
L_g	latent impact set of statement s_g
$x_{li}(s_g)$	latent impact factor of statement s_g
L_{g,t_k}^h	latent impact set of statement instance s_{g,t_k}^h
$x_{ih}(s_g)$	information hidden factor of statement s_g

References

1. Andrews, J.H.; Briand, L.C.; Labiche, Y. Is mutation an appropriate tool for testing experiments? In Proceedings of the 27th International Conference on Software Engineering, St. Louis, MO, USA, 15–21 May 2005; pp. 402–411.
2. DeMillo, R.A.; Lipton, R.J.; Sayward, F.G. Hints on test data selection: Help for the practicing programmer. *Computer* **1978**, *11*, 34–41. [CrossRef]
3. Mirshokraie, S.; Mesbah, A.; Pattabiraman, K. Efficient JavaScript mutation testing. In Proceedings of the 2013 IEEE Sixth International Conference on Software Testing, Verification and Validation, Luxembourg, 18–22 March 2013; pp. 74–83.
4. Jia, Y.; Harman, M. An analysis and survey of the development of mutation testing. *IEEE Trans. Softw. Eng.* **2011**, *37*, 649–678. [CrossRef]
5. Frankl, P.G.; Weiss, S.N.; Hu, C. All-uses vs mutation testing: An experimental comparison of effectiveness. *J. Syst. Softw.* **1997**, *38*, 235–253. [CrossRef]
6. Maldonado, J.C.; Delamaro, M.E.; Fabbri, S.C.; da Silva Simão, A.; Sugeta, T.; Vincenzi, A.M.R.; Masiero, P.C. Proteum: A family of tools to support specification and program testing based on mutation. In *Mutation Testing for the New Century*; Springer: Berlin/Heidelberg, Germany, 2001; pp. 113–116.
7. Acree, A.T., Jr. *On Mutation*; Technical Report; Georgia Inst of Tech Atlanta School of Information and Computer Science: Atlanta, GA, USA, 1980.
8. Zhang, L.; Hou, S.S.; Hu, J.J.; Xie, T.; Mei, H. Is operator-based mutant selection superior to random mutant selection? In Proceedings of the 32nd ACM/IEEE International Conference on Software Engineering, Cape Town, South Africa, 1–8 May 2010; Volume 1, pp. 435–444.
9. Hussain, S. Mutation Clustering. Master's Thesis, Kings College London, London, UK, 2008.

10. Gligoric, M.; Zhang, L.; Pereira, C.; Pokam, G. Selective mutation testing for concurrent code. In Proceedings of the 2013 International Symposium on Software Testing and Analysis, Lugano, Switzerland, 15–20 July 2013; pp. 224–234.
11. Offutt, A.J.; Rothermel, G.; Zapf, C. An experimental evaluation of selective mutation. In Proceedings of the 1993 15th International Conference on Software Engineering, Baltimore, MD, USA, 17–21 May 1993; pp. 100–107.
12. Zhang, J.; Zhang, L.; Harman, M.; Hao, D.; Jia, Y.; Zhang, L. Predictive mutation testing. *IEEE Trans. Softw. Eng.* **2018** . [CrossRef]
13. Jalbert, K.; Bradbury, J.S. Predicting mutation score using source code and test suite metrics. In Proceedings of the First, International Workshop on Realizing AI Synergies in Software Engineering, Zurich, Switzerland, 5 June 2012; pp. 42–46.
14. Goradia, T. Dynamic impact analysis: A cost-effective technique to enforce error-propagation. In Proceedings of the 1993 ACM SIGSOFT International Symposium on Software Testing and Analysis, Cambridge, MA, USA, 28–30 June 1993; Volume 18, pp. 171–181.
15. C programming Language Standard—C99. Available online: https://en.wikipedia.org/wiki/C99 (accessed on 19 August 2019).
16. The Program schedule.c. Available online: https://www.thecrazyprogrammer.com/2014/11/c-cpp-program-forpriority-scheduling-algorithm.html (accessed on 19 August 2019).
17. The Program tcas.c. Available online: https://sir.csc.ncsu.edu/php/showfiles.php (accessed on 19 August 2019).
18. Kuhn, M.; Johnson, K. *Applied Predictive Modeling*; Springer: Berlin/Heidelberg, Germany, 2013; Volume 26.
19. Brier, G.W. Verification of forecasts expressed in terms of probability. *Mon. Weather Rev.* **1950**, *78*, 1–3. [CrossRef]
20. Hosmer, D.W., Jr.; Lemeshow, S.; Sturdivant, R.X. *Applied Logistic Regression*; John Wiley & Sons: Hoboken, NJ, USA, 2013; Volume 398.
21. Harrell, F.E. *Regression Modeling Strategies*; Springer: Berlin/Heidelberg, Germany, 2001.
22. Hastie, T.; Tibshirani, R.; Wainwright, M. *Statistical Learning with Sparsity: The Lasso and Generalizations*; Chapman and Hall/CRC:Boca Raton, FL, USA, 2015.
23. Friedman, J.; Hastie, T.; Tibshirani, R. Regularization paths for generalized linear models via coordinate descent. *J. Stat. Softw.* **2010**, *33*, 1. [CrossRef] [PubMed]
24. Tibshirani, R. Regression Shrinkage and Selection Via the Lasso. *J. R. Stat. Soc. Ser. B* **1994**, *58*, 267–288. [CrossRef]
25. Glmnet. Available online: https://CRAN.R-project.org/package=glmnet (accessed on 19 August 2019).
26. Breiman, L. *Some Infinity Theory for Predictor Ensembles*; Technical Report 579; Statistics Dept. UCB: Berkeley, CA, USA, 2000.
27. Breiman, L. *Consistency for a Simple Model of Random Forests*; Technical Report (670); University of California at Berkeley: Berkeley, CA, USA, 2004.
28. Quinlan, J.R. Simplifying decision trees. *Int. J. Hum.-Comput. Stud.* **1999**, *51*, 497–510. [CrossRef]
29. The randomForest. Available online: https://cran.r-project.org/web/packages/randomForest/index.html (accessed on 19 August 2019).
30. Li, C. Probability Estimation in Random Forests. Master's Thesis, Department of Mathematics and Statistics, Utah State University, Logan, UT, USA, 2013.
31. Demuth, H.B.; Beale, M.H.; De Jess, O.; Hagan, M.T. *Neural Network Design*, 2nd ed.; Martin Hagan: Stillwater, OK, USA, 2014.
32. R Package nnet. Available online: https://CRAN.R-project.org/package=nnet (accessed on 19 August 2019).
33. Platt, J.C. Probabilistic Outputs for Support Vector Machines and Comparisons to Regularized Likelihood Methods. In *Advances in Large Margin Classifiers*; MIT Press: Cambridge, MA, USA, 1999; pp. 61–74.
34. Lin, H.T.; Lin, C.J.; Weng, R.C. A note on Platt's probabilistic outputs for support vector machines. *Mach. Learn.* **2007**, *68*, 267–276. [CrossRef]
35. Hsu, C.W.; Lin, C.J. A comparison of methods for multiclass support vector machines. *IEEE Trans. Neural Netw.* **2002**, *13*, 415–425. [PubMed]
36. Software Package Kernlab. Available online: https://CRAN.R-project.org/package=kernlab (accessed on 19 August 2019).

37. Caputo, B.; Sim, K.; Furesjo, F.; Smola, A. Appearance-based object recognition using SVMs: Which kernel should I use? In Proceedings of the NIPS Workshop on Statistical Methods for Computational Experiments in Visual Processing and Computer Vision, Whistler, BC, Canada, 12–14 December 2002; Volume 2002.
38. Karatzoglou, A.; Smola, A.; Hornik, K.; Zeileis, A. kernlab—An S4 package for kernel methods in R. *J. Stat. Softw.* **2004**, *11*, 1–20. [CrossRef]
39. The Symbolic Regression Tool Rgp. Available online: http://www.rdocumentation.org/packages/rg (accessed on 19 August 2019).
40. Sahoo, S.K.; Criswell, J.; Geigle, C.; Adve, V. Using likely invariants for automated software fault localization. In Proceedings of the 18th International Conference on Architectural Support for Programming Languages and Operating Systems, ASPLOS 2013, Houston, TX, USA, 16—20 March 2013; ACM SIGARCH Computer Architecture News; Volume 41, pp. 139–152.

 © 2019 by the authors. Licensee MDPI, Basel, Switzerland. This article is an open access article distributed under the terms and conditions of the Creative Commons Attribution (CC BY) license (http://creativecommons.org/licenses/by/4.0/).

Article

Extending the Characteristic Polynomial for Characterization of C_{20} Fullerene Congeners

Dan-Marian Joiţa [1] and Lorentz Jäntschi [1,2,*]

[1] Doctoral School of Chemistry, Babes-Bolyai University, 400028 Cluj, Romania; joita.danmarian@gmail.com
[2] Department of Physics and Chemistry, Technical University of Cluj-Napoca, 400641 Cluj, Romania
* Correspondence: lorentz.jantschi@gmail.com; Tel.: +40-264-401775

Received: 28 November 2017; Accepted: 13 December 2017; Published: 19 December 2017

Abstract: The characteristic polynomial (ChP) has found its use in the characterization of chemical compounds since Hückel's method of molecular orbitals. In order to discriminate the atoms of different elements and different bonds, an extension of the classical definition is required. The extending characteristic polynomial (EChP) family of structural descriptors is introduced in this article. Distinguishable atoms and bonds in the context of chemical structures are considered in the creation of the family of descriptors. The extension finds its uses in problems requiring discrimination among same-patterned graph representations of molecules as well as in problems involving relations between the structure and the properties of chemical compounds. The ability of the EChP to explain two properties, namely, area and volume, is analyzed on a sample of C_{20} fullerene congeners. The results have shown that the EChP-selected descriptors well explain the properties.

Keywords: characteristic polynomial (ChP); molecular descriptors; fullerene congeners; C_{20} fullerene; structure–property relationships

PACS: 02.10.Ox; 02.50.Sk; 02.50.Tt

MSC: 05C31; 12E10; 60E10; 55R40; 47N60

---

1. Introduction

The term 'secular function' has been used for what is now called a characteristic polynomial (ChP, in some of the literature, the term secular function is still used). The ChP was used to calculate secular perturbations (on a time scale of a century, i.e., slow compared with annual motion) of planetary orbits [1]. The first use of the ChP ($|\lambda \cdot Id - Ad|$, where Id is the identity matrix, and Ad is the adjacency matrix) in relation with chemical structure appeared after the discovery of wave-based treatment at the microscopic level [2]. The Hückel's method of molecular orbitals is actually the first extension of the ChP definition. He uses the 'secular determinant'—the determinant of a matrix which is decomposed as $|E \cdot Id - Ad|$, standing with the energy of the system (E instead of λ)—to approximate treatment of π electron systems in organic molecules [2].

The second extension of the ChP was found by Hartree [3,4] and Fock [5,6] by going in a different direction with the approximation of the wavefunction treatment. They actually found the same older eigenvector–eigenvalue problem (§20 in [7]; T1 in [8]) in Slater's treatment [9,10] of molecular orbitals. More generally (and older), the eigen-problem (finding of eigenvalues and eigenvectors) is involved in any Hessian [11] matrix [A] ([Ad] → [A], where Ad is the adjacency matrix). The Laplacian polynomial is a polynomial connected with the ChP (in Table 1). This uses a modified form (the Laplacian matrix, [La]) of the adjacency matrix ([Ad]), [La] = [Dg] − [Ad], where [Dg] simply counts on the main diagonal the number of the atom's bonds (the rest of its elements are null; for convenience with the graph-theory-related concept, it was denoted [Dg], from vertex degree). The

ChP is related also to the matching polynomial [12], degenerating to the same expression for forests (disjoint union of trees). Adapting [13] for molecules, a k-matching in a molecule is a matching with exactly k bonds between different atoms; see §3.1 & §3.3 in [14] for details. Each set containing a single edge is also an independent edge set; the empty set should be treated as an independent edge set with zero edges—this set is unique due to the constraint of connecting different atoms, where the matching may involve no more than $[n/2]$ bonds, where n is the number of atoms. It is possible to count the k-matches [15], but, nevertheless, it is a hard problem [16], as well as to express the derived Z-counting polynomial [17] and matching polynomial—both are defined using $m(k)$ as the k-matching number of the selected molecule, as shown in Table 1 (where n is the number of atoms).

Table 1. Characteristic polynomial (ChP), Laplacian polynomial (LaP), Z-counting, and Matching Polynomials.

Name	Formula
ChP	$\|\lambda \cdot [Id] - [Ad]\|$
LaP	$\|\lambda \cdot [Id] - [Dg] + [Ad]\|$
Z-counting	$\Sigma_{k \geq 0}\, m(k) \cdot \lambda^k$
Matching	$\Sigma_{k \geq 0}\, (-1)^k \cdot m(k) \cdot \lambda^{n-2k}$

A topological description of a molecule requires the storing of the bonds (as adjacencies) between the atoms and the atoms themselves (as identities). If this problem is simplified at maximum, by disregarding the atom and bond types, then the molecule is seen as an undirected and unweighted graph. The graph structure can be translated into the informational space by numbering the atoms. Unfortunately, this procedure also induces an isomorphism—the isomorphism of numbering, which may collapse into a nondeterministic polynomial time to be solved—see [18]. This is a reason for the desire of graph invariants, e.g., which do not depend on the numbering made on the graph.

Once the atoms (or the vertices) are numbered, the information can be simply stored as lists of vertices (V) and edges (E), and the graph structure of the molecule is associated with the pair G = (V, E). An equivalent representation is obtained using matrices. The adjacencies ([Ad]) are simply stored with 0 when no bond connects the atoms and 1 when a bond connecting the atoms exists. The identity matrix ([Id]) identifies the atoms by placing 1 on the main diagonal and 0 otherwise.

The ChP is the natural construction of a polynomial (in λ) in which the eigenvalues of [Ad] are the roots of the ChP as follows:

λ is an eigenvalue of [Ad] → there exists eigenvector $[v] \neq 0$ such that $\lambda \cdot [v] = [Ad] \times [v]$.

As a consequence:

$(\lambda \cdot [Id] - [Ad]) \cdot [v] = 0$; since $[v] \neq 0$ → $\lambda \cdot [Id] - [Ad]$ is singular → $|\lambda \cdot [Id] - [Ad]| = 0$.

Finally,

$$\text{ChP} \leftarrow |\lambda \cdot [Id] - [Ad]|.$$

ChP is a polynomial (in λ) of degree n, where n is the number of atoms. The ChP finds its uses in the topological theory of aromaticity [19,20], structure-resonance theory [21], quantum chemistry [22], and counts of random walks [23], as well as in eigenvector–eigenvalue problems [24].

This definition allows extensions. A natural extension is to store in the identity matrix ([Id]) non-unity instead of unity values ($[Id]_{i,j} = 1 \to [Id]_{i,j} \neq 1$) accounting for the atom types, as well as to store in the adjacency matrix ([Ad]) non-unity instead of unity values accounting for the bond types ($[Ad]_{i,j} = 1 \to [Ad]_{i,i} \neq 1$). This extension was subjected to study in the context of deriving structural descriptors useful for structure–property relationships.

2. Materials and Methods

2.1. Graphs, Matrices, and the Characteristic Polynomial

The topology of a graph structure could be expressed as matrices, and, in this regard, three of them are more frequently used: identity, adjacency (vertex–vertex, edge–edge, and vertex–edge), and distance matrices can be built (Table 2).

Table 2. Classical molecular graphs.

Definition	V: Finite Set	E ⊆ V × V	G = G(V,E)
Name (concept)	V: vertices (atoms)	E: edges (bonds)	G: graph (molecule)
Cardinality	|V| = n	|E| = m	$\forall n, V \leftrightarrow \{1, \ldots, n\}$
Example	G = "A-B-C"	V = {1,2,3}	E = {(1,2), (2,3)}

The matrices reflect in a 1:1 fashion the graph if the full graph is stored (each vertex pair stored twice, in both ways). The matrices of vertex adjacency ([Ad]) and of edge adjacency are square and the double enumeration of the edges is reflected in symmetry relative to the main diagonal (see Figure 1).

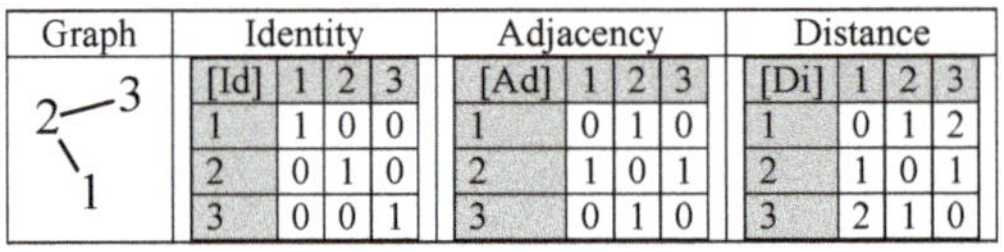

Graph	Identity				Adjacency				Distance			
2—3 \\ 1	[Id]	1	2	3	[Ad]	1	2	3	[Di]	1	2	3
	1	1	0	0	1	0	1	0	1	0	1	2
	2	0	1	0	2	1	0	1	2	1	0	1
	3	0	0	1	3	0	1	0	3	2	1	0

Figure 1. Encoded identities [I], adjacencies [A] and distances [D]—an example.

ChP is the natural construction of a polynomial in which the eigenvalues of the [Ad] are the roots of the ChP. ChP is a polynomial in λ of degree n, where n is the number of atoms. A natural extension is to store in [Id] (instead of unity) non-unity values accounting for the atom types, as well as to store in [Ad] (instead of unity) non-unity values accounting for the bond types.

An extremely important problem in chemistry is to uniquely identify a chemical compound. If the visual identification (looking at the structure) seems simple, for compounds of large size, this alternative is no longer viable. The data related to the structure of the compounds stored into the informational space may provide the answer to this problem. Nevertheless, together with the storing of the structure of the compound another issue is raised—namely, the arbitrary numbering of the atoms (Figure 2).

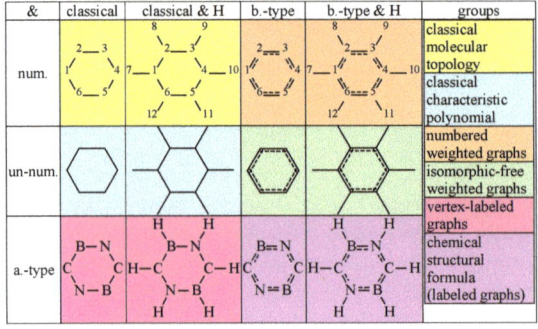

Figure 2. Graphs vs molecules—an example.

For a chemical structure with N atoms stored as a (classical molecular) graph, there exist exactly N! possibilities for numbering the atoms. Unfortunately, storing the graphs as lists of edges and (eventually) vertices does not provide a direct tool to check this arbitrary differentiation due to the numbering. The same situation applies to the adjacency matrices. Therefore, seeking for graph invariants is perfectly justified: an invariant (graph invariant) does not depend on numbering. The adjacency matrix is not a graph invariant and, therefore, it is necessary to go further than the adjacencies.

Important classes of graph invariants are the graph polynomials. To this category belongs the ChP—a graph invariant encoding important properties of the graph. On the other hand, unfortunately, ChP does not represent a bijective image of the graph, as there exist different graphs with the same ChP (i.e., cospectral graphs—the smallest cospectral graphs occurs for 5 vertices [25]). In order to count the cospectral graphs, one should compare A000088 and A082104 [26,27]. The ideal situation is that the invariant should be uniquely assigned to each structure, but this kind of invariant is difficult to find. A procedure to generate a non-degenerate invariant proposed by IUPAC is the international chemical identifier (InChI), which converts the chemical structure to a table of connectivity expressed as a unique and predictable series of characters [28].

Despite this inconvenience (not representing a bijective image of the graph) due to its link with the partition of the energy [2], the ChP seems to be one of the best alternatives for quantifying the information from the chemical structure.

Previously, researchers have shown the performance of estimation and/or prediction of the ChP on nonane isomers [29–31] as well as in the case of carbon nanostructures [32,33]. Furthermore, an online environment has been developed to assist researchers in the calculation of polynomials based on different approaches; this includes the ChP [34].

2.2. Characteristic Polynomial Extension

When doing calculations on molecular graphs, it is important to consider that, with the increase in the simplification in the graph representation (such as neglecting the type of the atom, bond orders, geometry in the favor of topology), the degeneration of the whole pool of possible calculations increases and there are more molecules with the same representation. This is favorable for the problems seeking similarities but is clearly unfavorable for the problems seeking dissimilarities.

A necessary step to accomplish better coverage of similarity vs dissimilarity dualism is to build and use a family of molecular descriptors, large enough to be able to provide answers for all. In the natural way, such a family should possess a 'genetic code'—namely, a series of variables from which to (re)produce a (one by one) molecular descriptor, all descriptors being therefore obtained in the same way. It is expected that all individuals of the family are independent of the numbering of the atoms in the molecule (should be molecular invariants).

The construction of such a family needs to consider the following:

- Molecules carry both topological and geometrical features (see Figure 3);
- Atom and bond types are essential factors in the expression of the measurable properties;
- Atom and/or bond numbering induces an undesired isomorphism;
- Geometry and bond types induce other kinds of isomorphism.

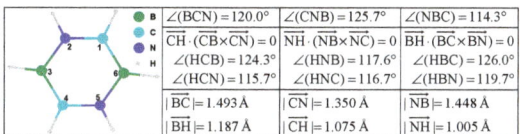

Figure 3. Molecular geometry—an example.

The representation of a molecule could be done using identity and adjacency (Figure 4).

[Ad]	1	2	3	4	5	6	→	[A]	1	2	3	4	5	6
1	0	1	0	0	0	1		1	0	a	0	0	0	c
2	1	0	1	0	0	0		2	a	0	b	0	0	0
3	0	1	0	1	0	0		3	0	b	0	c	0	0
4	0	0	1	0	1	0		4	0	0	c	0	a	0
5	0	0	0	1	0	1		5	0	0	0	a	0	b
6	1	0	0	0	1	0		6	c	0	0	0	b	0

[Id]	1	2	3	4	5	6	→	[I]	1	2	3	4	5	6
1	1	0	0	0	0	0		1	d	0	0	0	0	0
2	0	1	0	0	0	0		2	0	e	0	0	0	0
3	0	0	1	0	0	0		3	0	0	f	0	0	0
4	0	0	0	1	0	0		4	0	0	0	d	0	0
5	0	0	0	0	1	0		5	0	0	0	0	e	0
6	0	0	0	0	0	1		6	0	0	0	0	0	f

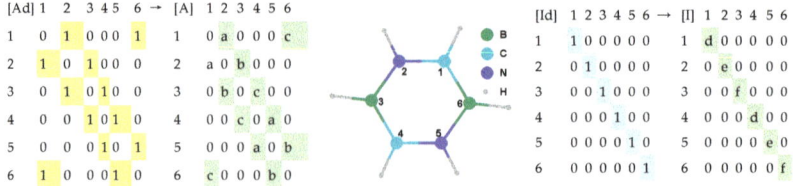

Figure 4. Molecular geometry translated into adjacency and identity—an example.

The distinct identities from Figure 4 are given using a, b, and c as variables in the case of adjacency and using d, e, and f as variables in the case of identity. This formalism allows the introduction of a natural extension of the ChP from graphs to molecules. There is no determinism in selecting the values of a–f. However,

- If a = b = c = d = e = f = 1 then ChPE ← ChP as in classical molecular topology.
- If a = b = c = 1.5^{-1}, then [A] accounts for the (inverse of the) bond order.
- If a = 1.35^{-1}, b = 1.448^{-1}, and c = 1.493^{-1} then [A] accounts for the (inverse of the) geometrical distance (in Å).
- If d = 12/294, e = 14/294, and f = 10.8/294, then [I] accounts for atomic mass relative to Uuo, the last element from the 7th period of the system of elements.
- If d = 2267/ρ_{ref}, e = 1026/ρ_{ref}, and f = 2460/ρ_{ref}, then [I] accounts for the solid state relative density (in m^3/kg); ρ_{ref} can be fixed to 30,000.
- If d = 2.55/4.00, e = 3.04/4.00, and f = 2.04/4.00, then [I] accounts for electronegativity relative to Fluorine when the Pauling scale is used.
- If d = 1086.2/1312, e = 1402.3/1312, and f = 800.6/1312, then [I] accounts for the first potential of ionization relative to the potential of ionization for Hydrogen.
- If d = 3820/3820, e = 63/3820, and f = 2573/3820, then [I] accounts for melting point relative to the diamond allotrope of Carbon (in K).
- If d = 1/4, e = 1/4, and f = 1/4, then [I] accounts for the number of hydrogen atoms attached relative to the score of CH_4.

The full extension could include also the distance matrix (Figure 5).

[Ad]	1	2	3	4	5	6	[Id]	1	2	3	4	5	6		[Di]	1	2	3	4	5	6	[Id]	1	2	3	4	5	6
1	0	1	0	0	0	1	1	1	0	0	0	0	0	→	1	0	1	2	3	2	1	1	1	0	0	0	0	0
2	1	0	1	0	0	0	2	0	1	0	0	0	0	extension	2	1	0	1	2	3	2	2	0	1	0	0	0	0
3	0	1	0	1	0	0	3	0	0	1	0	0	0		3	2	1	0	1	2	3	3	0	0	1	0	0	0
4	0	0	1	0	1	0	4	0	0	0	1	0	0		4	3	2	1	0	1	2	4	0	0	0	1	0	0
5	0	0	0	1	0	1	5	0	0	0	0	1	0	→	5	2	3	2	1	0	1	5	0	0	0	0	1	0
6	1	0	0	0	1	0	6	0	0	0	0	0	1		6	1	2	3	2	1	0	6	0	0	0	0	0	1
↓ extension ↓															↓ extension ↓													
[A]	1	2	3	4	5	6	[I]	1	2	3	4	5	6		[D]	1	2	3	4	5	6	[I]	1	2	3	4	5	6
1	0	a	0	0	0	c	1	d	0	0	0	0	0	→	1	0	a	h	k	g	c	1	d	0	0	0	0	0
2	a	0	b	0	0	0	2	0	e	0	0	0	0	extension	2	a	0	b	g	j	i	2	0	e	0	0	0	0
3	0	b	0	c	0	0	3	0	0	f	0	0	0		3	h	b	0	c	i	l	3	0	0	f	0	0	0
4	0	0	c	0	a	0	4	0	0	0	d	0	0		4	k	g	c	0	a	h	4	0	0	0	d	0	0
5	0	0	0	a	0	b	5	0	0	0	0	e	0	→	5	g	j	i	a	0	b	5	0	0	0	0	e	0
6	c	0	0	0	b	0	6	0	0	0	0	0	f		6	c	i	l	h	b	0	6	0	0	0	0	0	f

Figure 5. Molecular geometry translated into adjacency, identity, and distance—an example.

The extended ChP has the following formula:

$$ChP \leftarrow |\lambda \times [I] - [C]|$$

where [C] is either [A] or [D], the identities (a, b, and c from [I]) and the connectivity (d, e, f, g, h, i, j, k, and l from [C]).

The single-value entries (0 and $1 \neq 0$ for the classical definition of the ChP) can be upgraded to multi-value (any value), accounting for different atoms and bonds. Obviously, the classical ChP is found when a = b = c = d = e = f = 1 and g = h = i = j = k = l = 0.

Figure 6 shows the ChP extension differently accounting the identities from atomic properties ($[I] \leftarrow A_P \in \{A, B, C, D, E, F, G, H, I, J, K, L\}$) and connectivity properties ($[C] \leftarrow C_P \in \{t, g, c, b, T, G, C, B,\}$).

A_P	Property	A_P	Property	A_P	Property	A_P	Property
A	Atomic mass	D	Density	G	Melting point	J	Mulliken charge
B	Boiling point	E	Electronegativity	H	Hydrogen connections	K	Natural charge
C	Count	F	First ionization potential	I	Electrostatic charge	L	Spin

$$\text{ChPE} \overset{\text{def}}{=} |\lambda \cdot I - C|,\ [I] \leftarrow \text{identity properties } (\uparrow),\ [C] \leftarrow \text{connectivity properties } (\downarrow)$$

C_P	from adjacencies	C_P	from all connections	Parameters
t	[C] ← [Ad] (classical ChP)	T	[C] ← [Di] ([Ad] ← [Di])	topological connections
g	'1' ← (geometrical distance)$^{-1}$	G	'1' ← (geometrical distance)$^{-1}$	geometrical coordinates
c	'1' ← (bond order)$^{-1}$	C	'1' ← (bond orders sum)$^{-1}$	conventional bond orders
b	'1' ← (bond order)$^{-1}$	B	'1' ← (bond orders sum)$^{-1}$	Mulliken bond orders

$$\text{ChPE}(\lambda, I_P, C_P) \rightarrow L_L L_I L_C(\pm d_0.d_1d_2d_3)$$

Linearization $L_L \in \{I, R, L\}$, $f_I(x) = x$, $f_R(x) = x^{-1}$, $f_L(x) = \ln(x)$
Identity $L_I \in \{A, \ldots, L\}$, Connectivity $L_C \in \{t, g, c, b, T, G, C, B\}$
Evaluation $d_0 \in \{0,1\}$, $d_1, d_2, d_3 \in \{0, \ldots, 9\}$ ($\lambda = \pm d_0.d_1d_2d_3$, 2001 evaluation points)

Figure 6. Extended characteristic polynomial—EChP.

The extending characteristic polynomial (EChP) is designed for estimation/prediction of molecular properties, so a software implementation was done. EChP(λ, I_P, C_P) diverges as ChP(λ) does (to ∞) quickly with the increase of $\lambda > 1$. Thus, the $[-1, 1]$ range $\rightarrow$ '2001' grid is useful for evaluation. A linearization (L_L) is required and was implemented since biological properties are expressed in log scale. The evaluation is performed at every point (out of 2001), requiring $O(n^3)$ operations (where n is the number of atoms).

EChP is a family with 96 (n_I*n_C) polynomial formulas and 288 (*n_L) linearized ones, leading to a total of 576,288 individuals. The FreePascal software was used for implementation since it is very fast and allows a parallelized version to be used with multi-CPUs (chp17chp.pas) [35]. The program requires input files in the 'chp' format (such as chfp_17_q.asc, see Figure 7), and uses a filtering (PHP) program ($\rightarrow$chfp_17_t.asc) as well as a molecular property file (such as chfp_17 [prop].txt). The filtering program was designed to look for degenerations and to reduce the pool of descriptors by eliminating the degenerated ones.

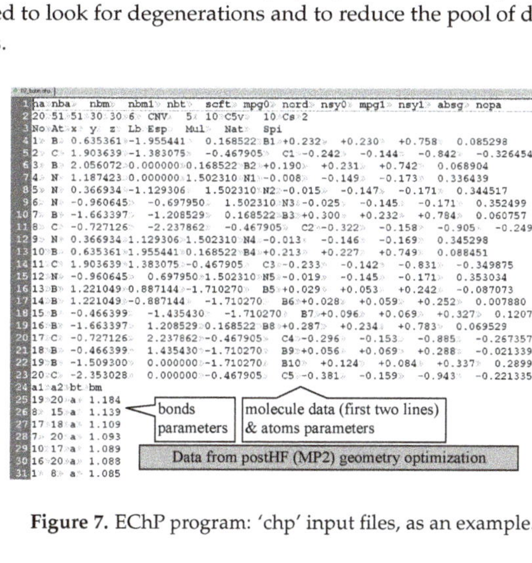
```
na nba nbm nbm1 nbt scft mpg0 nord nsy0 mpg1 nsy1 absg nopa
20 51 51 30 30 6 CNV 5 10 C5v 10 Cs 2
No At x y z Lb Esp Mul Nat Spi
1 B 0.635361 -1.955441 0.168522 B1 +0.232 +0.230 +0.758 0.085298
2 C 1.903639 -1.383075 -0.467905 C1 -0.242 -0.144 -0.842 -0.326454
3 B 2.056072 0.000000 0.168522 B2 +0.190 +0.231 +0.742 0.068904
4 N 1.187423 0.000000 1.502310 N1 -0.008 -0.149 -0.173 0.336439
5 N 0.366934 -1.129306 1.502310 N2 -0.015 -0.147 -0.171 0.344517
6 N -0.960645 -0.697950 1.502310 N3 -0.025 -0.145 -0.171 0.352499
7 B -1.663397 -1.208529 0.168522 B3 +0.300 +0.232 +0.784 0.060757
8 C -0.727126 -2.237862 -0.467905 C2 -0.322 -0.158 -0.905 -0.249847
9 N 0.366934 1.129306 1.502310 N4 -0.013 -0.146 -0.169 0.345298
10 B 0.635361 1.955441 0.168522 B4 +0.213 +0.227 +0.749 0.088451
11 C 1.903639 1.383075 -0.467905 C3 -0.233 -0.142 -0.831 -0.349875
12 N -0.960645 0.697950 1.502310 N5 -0.019 -0.145 -0.171 0.353034
13 B 1.221049 0.887144 -1.710270 B5 +0.029 +0.053 +0.242 -0.087073
14 B 1.221049 -0.887144 -1.710270 B6 +0.028 +0.059 +0.252 0.007880
15 B -0.466399 -1.435430 -1.710270 B7 +0.096 +0.069 +0.327 0.120770
16 B -1.663397 1.208529 0.168522 B8 +0.287 +0.234 +0.783 0.069529
17 C -0.727126 2.237862 -0.467905 C4 -0.296 -0.153 -0.885 -0.267357
18 B -0.466399 1.435430 -1.710270 B9 +0.056 +0.069 +0.288 -0.021339
19 B -1.509300 0.000000 -1.710270 B10 +0.124 +0.084 +0.337 0.289904
20 C -2.353028 0.000000 -0.467905 C5 -0.381 -0.159 -0.943 -0.221335
a1 a2 bt bm
19 20 a 1.184
8 15 a 1.139
17 18 a 1.109
7 20 a 1.093
10 17 a 1.089
16 20 a 1.088
1 8 a 1.085
```


Figure 7. EChP program: 'chp' input files, as an example.

The family of EChP descriptors was then used with a series of chemical compounds to obtain associations between the structure and properties as regression equations.

2.3. Numerical Case Study

The case study was conducted on C_{20} fullerene congeners with Boron, Carbon, or Nitrogen atoms on each layer (Figure 8). A sample of 45 distinct compounds was obtained. The generic name of the files was stored as dd_$R_1R_2R_3R_4$, where dd is the number of the compound in the set and R_1–R_4 are the atoms on layers 1–4 (e.g., 02_bbbn.chp is the second compound in the sample and has boron of the first three layers and nitrogen on the last layer).

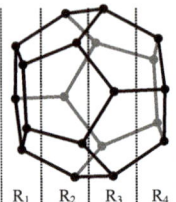

Figure 8. C_{20} fullerene congeners (R is the symbol of the atom on the layer).

The geometries were built at the Hartree-Fock (HF) [3–6] 6-31 G [36] level of theory and calculated properties (namely, area and volume) were extracted from these calculations. Two different structures proved stable for bbbb (see Figure 9) and both were included in the analysis, resulting in a sample of 46 compounds.

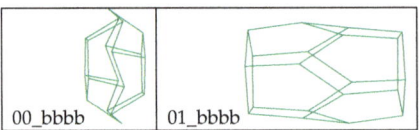

Figure 9. bbbb C_{20} stable fullerenes.

The values of the calculated properties are given in Table 3.

Table 3. C_{20} congeners: values of investigated properties.

Mol	Area	Volume	Mol	Area	Volume	Mol	Area	Volume
00_bbbb	54.641	30.063	16_cbbb	50.537	27.863	31_ccnc	42.689	22.542
01_bbbb	51.863	26.948	17_cbbc	51.114	29.107	32_ccnn	43.987	23.862
02_bbbn	54.848	32.333	18_cbbn	49.097	27.424	33_cnbb	49.186	28.569
03_bbcn	48.481	27.524	19_cbcb	51.733	30.156	34_cnbn	44.694	24.794
04_bbnb	53.093	30.658	20_cbcn	47.401	26.543	35_cncb	46.994	26.275
05_bbnn	49.797	27.573	21_cbnb	48.262	26.68	36_cncn	44.723	24.062
06_bcbb	54.597	32.043	22_cbnc	45.944	25.109	37_cnnb	45.76	24.995
07_bcbn	49.415	28.726	23_cbnn	45.578	24.689	38_cnnc	48.834	24.315
08_bccb	51.676	29.739	24_ccbb	52.365	30.954	39_cnnn	45.508	24.847
09_bccn	47.392	26.933	25_ccbc	45.618	24.718	40_nbbn	48.119	26.881
10_bcnb	48.782	26.786	26_ccbn	45.857	25.514	41_nbnn	45.726	24.275
11_bcnn	47.15	25.543	27_cccb	46.446	25.49	42_ncbn	45.735	25.533
12_bnbn	47.791	27.383	28_cccc	43.707	23.584	43_nccn	45.211	24.676
13_bncn	47.048	26.368	29_cccn	43.86	23.926	44_ncnn	44.848	24.445
14_bnnb	48.244	27.25	30_ccnb	45.901	25.525	45_nnnn	46.463	25.872
15_bnnn	47.226	25.93	-	-	-	-	-	-

Normal distribution of the data is one assumption that needs to be assessed before any linear regression analysis. Six different tests were used (AD = Anderson-Darling, KS = Kolmogorov-Smirnov, CM = Cramér-von Mises, KV = Kuiper V, WU = Watson U^2, H1 = Shannon's entropy [37]) [38] and the decision was made based on the combined test proposed by Fisher [39]. The distribution of the investigated properties proved to be not significantly different from the expected normal distribution (see Table 4, all p-values > 0.05).

Table 4. C_{20} congeners: values of investigated properties. AD = Anderson–Darling; KS = Kolmogorov–Smirnov; CM = Cramér–von Mises; KV = Kuiper V; WU = Watson U2; H1 = Shannon's entropy.

Prop.	Title	AD	KS	CM	KV	WU	H1	FCS(6)
area	stat	0.826	0.758	0.131	1.213	0.110	22.83	3.660
	p	0.462	0.423	0.548	0.552	0.770	0.565	0.723
volume	stat	0.845	0.791	0.133	1.272	0.108	22.95	3.503
	p	0.445	0.477	0.552	0.633	0.765	0.525	0.744

Where for a series of cumulative distribution function values $((f_i)_{1 \leq i \leq n})$:

Statistic	Formula
AD	$-n - \frac{1}{n}\sum_{i=1}^{n}(2 \cdot i - 1) \cdot \ln(f_i \cdot (1 - f_{n+1-i}))$
KS	$\sqrt{n} \cdot \max_{1 \leq i \leq n}\left(f_i - \frac{i-1}{n}, \frac{i}{n} - f_i\right)$
CM	$\frac{1}{12n} + \sum_{i=1}^{n}\left(\frac{2 \cdot i - 1}{2n} - f_i\right)^2$
KV	$\sqrt{n} \cdot \left(\max_{1 \leq i \leq n}\left(f_i - \frac{i-1}{n}\right) + \max_{1 \leq i \leq n}\left(\frac{i}{n} - f_i\right)\right)$
WU	$CM - n\left(\frac{1}{n}\sum_{i=1}^{n}f_i - \frac{1}{2}\right)^2$
H1	$-\sum_{i=1}^{n}f_i \cdot \ln(f_i) - \sum_{i=1}^{n}(1 - f_i) \cdot \ln(1 - f_i)$
FCS	$\ln(p_{AD} \cdot p_{KS} \cdot p_{CM} \cdot p_{KV} \cdot p_{WU} \cdot p_{H1})$

The absences of the outliers have also been investigated using Grubb's test [40] for the association between volume (vol) and area on the sample of investigated C_{20} congeners. The analysis identified three compounds as outliers, their exclusion leading to a performing linear association (Figure 10).

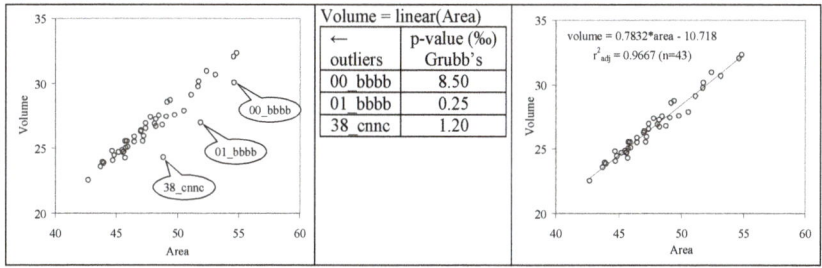

Figure 10. Volume as linear function of area.

The values of the EChP descriptors were generated for all molecules in the dataset and were used as input data for searching linear regression models able to explain the investigated properties (area and volume). Three different approaches were used, searching for additive, multiplicative, or full linear dependence (see Table 5).

Table 5. Approaches in bivariate ($k_D = 2$) regression analysis.

	$Y \sim \hat{Y} = a_0 + a_1*ChPE_1 + a_2*ChPE_2 + a_3*ChPE_1*ChPE_2$	
Effect	Coefficient Constraints	k_C
Additive ("+")	$a_0 = 0, a_1 \neq 0, a_2 \neq 0, a_3 = 0$	2 (a_1, a_2)
	$a_0 \neq 0, a_1 \neq 0, a_2 \neq 0, a_3 = 0$	3 (a_0, a_1, a_2)
Multiplicative ("*")	$a_0 = 0, a_1 = 0, a_2 = 0, a_3 \neq 0$	1 (a_3)
	$a_0 \neq 0, a_1 = 0, a_2 = 0, a_3 \neq 0$	2 (a_0, a_3)
Full	$a_0 = 0, a_1 \neq 0, a_2 \neq 0, a_3 \neq 0$	3 (a_1, a_2, a_3)
	$a_0 \neq 0, a_1 \neq 0, a_2 \neq 0, a_3 \neq 0$	4 (a_0, a_1, a_2, a_3)

The selection of the performing models was done using the adjusted determination coefficient ($r^2_{adj} = r^2 - (1 - r^2)*k_D*(n - k_C)^{-1}$, where n is the number of compounds in the model). The difference between models with the same properties was tested using the studentized version of the Fisher Z transformation [41,42].

The best-performing models identified for the investigated properties are presented in Table 6 while the characteristics of the models are given in Table 7.

Table 6. ChPE models.

Eff	P	Model	eq
"+"	A	$35.8_{\pm 0.3} - 8.2_{\pm 0.1} * LCG_{+0.238} + 1.4_{\pm 0.3} * LCG_{-0.896}$	1 [a]
	V	$21.6_{\pm 2.0} - 7.4_{\pm 0.7} * LCG_{+0.238} + 1.7_{\pm 0.3} * LCG_{-0.896}$	2
"*"	A	$34.0_{\pm 0.9} + 0.16_{\pm 0.01} * LEG_{+0.436} * LFG_{-0.952}$	3
	V	$17.6_{\pm 1.0} + 0.101_{\pm 0.011} * LEG_{+0.436} * LCG_{-0.384}$	4
Full	A	$50.4_{\pm 0.5} - 6.36_{\pm 0.06} * LCG_{+0.276} + 2.3_{\pm 0.5} * LCG_{-0.908} + 0.13_{\pm 0.06} * LCG_{+0.276} * LCG_{-0.908}$	5
	V	$64_{\pm 17} - 2.5_{\pm 1.9} * LCG_{+0.236} + 4.5_{\pm 1.2} * LCG_{-0.908} + 0.35_{\pm 0.14} * LCG_{+0.236} * LCG_{-0.908}$	6

Eff = Effect; "+" = additive model; "*" = multiplicative model; P = property: A = Area, V = Volume. [a] 03_bbcn excluded outlier.

Table 7. Model characteristics.

Eff	P	eq	r^2_{adj}	se	F (p-Value)
"+"	A	1	0.9934	0.2487	3386 (5.01 × 10^{-48})
	V	2	0.9385	0.5767	344 (3.41 × 10^{-27})
"*"	A	3	0.9462	0.6575	931 (3.06 × 10^{-31})
	V	4	0.8894	0.7651	372 (4.37 × 10^{-23})
Full	A	5	0.9940	0.2406	2413 (5.04 × 10^{-47})
	V	6	0.9462	0.5458	258 (4.37 × 10^{-27})

Eff = Effect; "+" = additive model; "*" = multiplicative model; P = property: A = Area, V = Volume, r^2_{adj} = adjusted determination coefficient; se = standard error of estimate, F (p-value) = Fisher's statistic (associated significance).

The relationship between volume and area is translated in the identification of the same EChP descriptors as the explanatory variable (two descriptors for additive models and one descriptor for multiplicative and respective full model, see Table 6). All models had a capacity of explanation higher than 85%, with the worst performance obtained by multiplicative models and similar performances (without significant difference) obtained by additive and full models (see Table 8).

Table 8. Fisher's Z model comparisons: results.

Prop.	Parameter	"*" vs "+"	"*" vs Full	"+" vs Full
Area	Stat	4.61	4.82	0.21
	p-value	<0.0001	<0.0001	0.4176
Volume	Stat	1.42	1.74	0.32
	p-value	0.0791	0.0425	0.3752

Graphical representations of calculated and estimated area and respective volume by the investigated effects are given in Figure 11 (eq1–eq3) and Figure 12 (eq4–eq6).

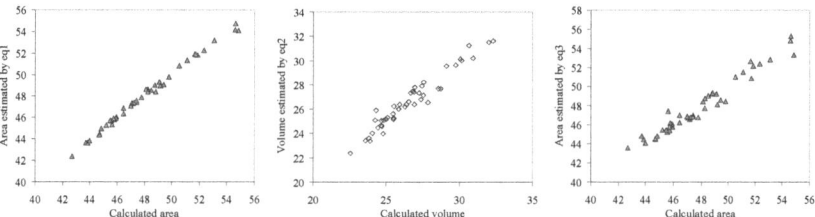

Figure 11. Graphical representation of eq1–eq3 model performances.

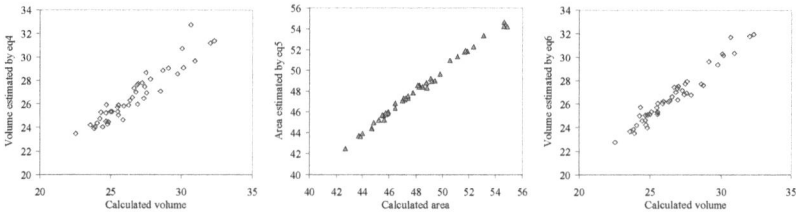

Figure 12. Graphical representation of eq4–eq6 model performances.

The model comparison strongly suggests that the best performing models are the additive or the full model for both investigated properties. However, since 03_bbcn is an outlier for the area on the additive model, we can say that choosing the full model will give a correct estimation.

It is important that the performing models identified using the EChP descriptors—the full model—select the same polynomial for both descriptors when both area and volume ("CG" in $LCG_{+0.236}$, $LCG_{+0.276}$, and $LCG_{-0.908}$) are investigated. It should be noted that one descriptor is common for the estimation of the area and of the volume ($LCG_{-0.908}$) for the C_{20} fullerene congeners. This fact, in conjunction with the higher correlation between volume and area ($r^2_{adj} \approx 0.97$), the presence of outliers in one additive model, and the significant higher performance by full models in estimation sustained by goodness-of-fit and the graphical representation of calculated versus estimated, suggests that the best models are those with full effects.

3. Conclusions and Further Work

EChP proved useful for estimation of the investigated molecular properties. Both properties of C_{20} congeners—volume and area—are explained by a common descriptor ($LCG_{-0.908}$ (or vice versa)).

EChP is a natural extension of the ChP. The scales of the atomic properties were more or less arbitrary selected and will be further investigated to find the optimal solution. Furthermore, the reversed distance seemed to be the best alternative but further analysis must be conducted to demonstrate this observation.

Author Contributions: Dan-Marian Joiţa made the molecules and supervised the molecular geometry optimization (energy minimization). Lorentz Jäntschi supervised the whole study and wrote the paper.

Conflicts of Interest: The authors declare no conflict of interest.

References

1. Lagrange, J.-L. *Sur L'équation Séculaire de la Lune*; Mémoires de l'Acadéémie Royale des Science: Paris, France, 1773.
2. Huckel, E. Quantentheoretische Beiträge zum Benzolproblem. *Z. Phys.* **1931**, *70*, 204–286. [CrossRef]
3. Hartree, D.R. The Wave Mechanics of an Atom with a Non-Coulomb Central Field. Part I. Theory and Methods. *Math. Proc. Camb. Philos. Soc.* **1928**, *24*, 89–110. [CrossRef]
4. Hartree, D.R. The Wave Mechanics of an Atom with a Non-Coulomb Central Field. Part II. Some Results and Discussion. *Math. Proc. Camb. Philos. Soc.* **1928**, *24*, 111–132. [CrossRef]
5. Fock, V.A. Näherungsmethode zur Lösung des quantenmechanischen Mehrkörperproblems. *Z. Phys.* **1930**, *61*, 26–148. [CrossRef]
6. Fock, V.A. "Selfconsistent field" mit Austausch für Natrium. *Z. Phys.* **1930**, *62*, 795–805. [CrossRef]
7. Laplace, P.S. *Recherches sur le Calcul Intégral et sur le Système du Monde*; Mémoires l'Académie des Sciences: Paris, France, 1776; Volume 2, pp. 47–179.
8. Cauchy, A. Sur l'équation à l'aide de laquelle on détermine les inégalités séculaires des mouvements des planets. *Exerc. Math.* **1829**, *4*, 140–160.
9. Slater, J.C. The Theory of Complex Spectra. *Phys. Rev.* **1929**, *34*, 1293–1295. [CrossRef]
10. Hartree, D.R.; Hartree, W. Self-Consistent Field, with Exchange, for Beryllium. *Proc. R. Soc. A Math. Phys. Eng. Sci.* **1935**, *50*, 9–33. [CrossRef]
11. Sylvester, J.J. On the theory connected with Newton's rule for the discovery of imaginary roots of equations. *Messenger Math.* **1880**, *9*, 71–84.
12. Godsil, C.D.; Gutman, I. On the theory of the matching polynomial. *J. Graph Theory* **1981**, *5*, 137–144. [CrossRef]
13. Godsil, C.D. Algebraic Matching Theory. *Electron. J. Comb.* **1995**, *2*, #R8.
14. Diudea, M.V.; Gutman, I.; Jäntschi, L. *Molecular Topology*; Nova Science: New York, NY, USA, 2001.
15. Ramaraj, R.; Balasubramanian, K. Computer generation of matching polynimials of chemical graphs and lattices. *J. Comput. Chem.* **1985**, *6*, 122–141. [CrossRef]
16. Curticapean, R. Counting Matchings of Size k Is # W[1]-Hard. In Proceedings of the 40th International Conference on Automata, Languages, and Programming, ICALP'13, Riga, Latvia, 8–12 July 2013; Volume 7965, pp. 352–363.
17. Hosoya, H. Topological Index. A Newly Proposed Quantity Characterizing the Topological Nature of Structural Isomers of Saturated Hydrocarbons. *Bull. Chem. Soc. Jpn.* **1971**, *44*, 2332–2339. [CrossRef]
18. Schöning, U. Graph isomorphism is in the low hierarchy. *J. Comput. Syst. Sci.* **1987**, *37*, 312–323. [CrossRef]
19. King, R.B. Applications of graph theory and topology for the study of aromaticity in inorganic compounds. *J. Chem. Inf. Model.* **1992**, *32*, 42–47. [CrossRef]
20. Santos, J.C.; Andres, J.; Aizman, A.; Fuentealba, P. An Aromaticity Scale Based on the Topological Analysis of the Electron Localization Function Including σ and π Contributions. *J. Chem. Theory Comput.* **2005**, *1*, 83–86. [CrossRef] [PubMed]
21. Herndon, W.C. Structure-resonance theory for pericyclic transition states. *J. Chem. Educ.* **1981**, *58*, 371. [CrossRef]
22. Bruderer, M.; Contreras-Pulido, L.D.; Thaller, M.; Sironi, L.; Obreschkow, D.; Plenio, M.B. Inverse counting statistics for stochastic and open quantum systems: The characteristic polynomial approach. *New J. Phys.* **2014**, *16*, 033030. [CrossRef]
23. Arguin, L.-P.; Belius, D.; Bourgade, P. Maximum of the Characteristic Polynomial of Random Unitary Matrices. *Commun. Math. Phys.* **2017**, *349*, 703–751. [CrossRef]
24. Da Lita Silva, J. On the characteristic polynomial, eigenvectors and determinant of heptadiagonal matrices. *Linear Multilinear Algebra* **2017**, *65*, 1852–1866. [CrossRef]
25. Collatz, L.; Sinogowitz, U. Spektren Endlicher Grafen. *Abh. Math. Semin. Univ. Hambg.* **1957**, *21*, 63–77. [CrossRef]

26. Sloane, N.J.A. *Number of Graphs on n Unlabeled Nodes*; A000088; On-Line Encyclopedia of Integer Sequences (OEIS): Highland Park, NJ, USA, 1996.
27. Weisstein, W.E. *Number of Unique Characteristic Polynomials among All Simple Undirected Graphs on n Nodes*; A082104; On-Line Encyclopedia of Integer Sequences (OEIS): Highland Park, NJ, USA, 2003.
28. McNaught, A. The IUPAC international chemical identifier. *Chem. Int.* **2006**, *28*, 12–15.
29. Jäntschi, L.; Bolboacă, S.D.; Furdui, C.M. Characteristic and counting polynomials: Modelling nonane isomers properties. *Mol. Simul.* **2009**, *35*, 220–227. [CrossRef]
30. Bolboacă, S.D.; Jäntschi, L. How good can the characteristic polynomial be for correlations? *Int. J. Mol. Sci.* **2007**, *8*, 335–345. [CrossRef]
31. Jäntschi, L. *Characteristic and Counting Polynomials of Nonane Isomers*; Academic Direct Publishing House: Cluj-Napoca, Romania, 2007; ISBN 978-973-86211-3-8.
32. Bolboacă, S.D.; Jäntschi, L. Characteristic Polynomial in Assessment of Carbon-Nano Structures. In *Sustainable Nanosystems Development, Properties, and Applications*; Putz, M.V., Mirica, M.C., Eds.; IGI Global: Hershey, PA, USA, 2017; pp. 122–147, ISBN 9781522504924.
33. Bolboacă, S.D.; Jäntschi, L. Counting Distance and Szeged (on Distance) Polynomials in Dodecahedron Nano-assemblies. In *Distance, Symmetry, and Topology in Carbon Nanomaterials*; Ashrafi, A.R., Diudea, M.V., Eds.; Springer International Publishing: Cham, Switzerland, 2016; pp. 391–408, ISBN 978-3-319-31582-9.
34. Jäntschi, L. Online Calculation of Graph Polynomials Such as Counting Polynomial and Characteristic Polynomial. 2006. Available online: http://l.academicdirect.org/Fundamentals/Graphs/polynomials/ (accessed on 21 January 2017).
35. Gabor, B.M.; Vreman, P.P. Free Pascal: Open Source Compiler for Pascal and Object Pascal. 1988 (and to Date). Available online: http://freepascal.org (accessed on 21 January 2017).
36. Hehre, W.J.; Ditchfield, R.; Pople, J.A. Self-consistent molecular orbital methods. XII. Further extensions of Gaussian-type basis sets for use in molecular orbital studies of organic molecules. *J. Chem. Phys.* **1972**, *56*, 2257–2261. [CrossRef]
37. Jäntschi, L.; Bolboacă, S.D. Performances of Shannon's Entropy Statistic in Assessment of Distribution of Data. *Ovidius Univ. Ann. Chem.* **2017**, *28*, 30–42. [CrossRef]
38. Jäntschi, L. Tests. Available online: http://l.academicdirect.ro/Statistics/tests/ (accessed on 1 March 2017).
39. Fisher, R.A. Questions and answers #14. *Am. Stat.* **1948**, *2*, 30–31.
40. Bolboacă, S.D.; Jäntschi, L. Distribution Fitting 3. Analysis under Normality Assumptions. *Bull. Univ. Agric. Sci. Vet. Med. Cluj-Napoca. Hortic.* **2009**, *66*, 698–705.
41. Student. The probable error of a mean. *Biometrika* **1908**, *6*, 1–25. [CrossRef]
42. Welch, B.L. The generalization of student's problem when several different population variances are involved. *Biometrika* **1947**, *34*, 28–35. [CrossRef] [PubMed]

© 2017 by the authors. Licensee MDPI, Basel, Switzerland. This article is an open access article distributed under the terms and conditions of the Creative Commons Attribution (CC BY) license (http://creativecommons.org/licenses/by/4.0/).

MDPI
St. Alban-Anlage 66
4052 Basel
Switzerland
Tel. +41 61 683 77 34
Fax +41 61 302 89 18
www.mdpi.com

Mathematics Editorial Office
E-mail: mathematics@mdpi.com
www.mdpi.com/journal/mathematics

www.ingramcontent.com/pod-product-compliance
Lightning Source LLC
LaVergne TN
LVHW072000080526
838202LV00064B/6798